Flash! The hunt for the biggest explosions in the universe

The origin and nature of gamma ray bursts is currently one of the greatest mysteries in astrophysics. These tremendously powerful blasts produce more energy in a fraction of a second than our sun does in ten billion years. Since their accidental discovery by American spy satellites over thirty years ago, astronomers have striven to understand these enigmatic explosions. It is only recently, thanks to an Italian-Dutch satellite, and powerful telescopes both on the ground and in space, that the mystery is beginning to be unravelled. Astronomers now realize that gamma ray bursts are probably related to the birth of black holes in extremely distant galaxies. *Flash!* describes the fast-moving field of gamma ray burst research, from the initial discovery right up to the most recent developments. Based on interviews with leading scientists, this exciting book provides an inside view of the scientific challenges involved in unravelling the mystery of gamma ray bursts.

GOVERT SCHILLING is a Dutch science writer and astronomy publicist. He is contributing editor of *Sky & Telescope* magazine, and regularly writes for the news sections of *Science* and *New Scientist*. Schilling is the astronomy writer for *de Volkskrant*, one of the largest national daily newspapers in the Netherlands, and frequently talks about the universe on Dutch radio broadcasts. He is author of more than twenty popular astronomy books, and hundreds of newspaper and magazine articles on astronomy.

Flash!

The hunt for the biggest explosions in the universe

Govert Schilling

Translated by
Naomi Greenberg-Slovin

CAMBRIDGE
UNIVERSITY PRESS

PUBLISHED BY THE PRESS SYNDICATE OF THE UNIVERSITY OF CAMBRIDGE
The Pitt Building, Trumpington Street, Cambridge, United Kingdom

CAMBRIDGE UNIVERSITY PRESS
The Edinburgh Building, Cambridge CB2 2RU, UK
40 West, 20th Street, New York, NY 10011–4211, USA
477 Williamstown Road, Port Melbourne, VIC 3207, Australia
Ruiz de Alarcón 13, 28014 Madrid, Spain
Dock House, The Waterfront, Cape Town 8001, South Africa

http://www.cambridge.org

First published in Dutch by Uitgeverij Wereldbibliotheek, Amsterdam, 2000
English edition published 2002

Printed in the United Kingdom at the University Press, Cambridge

Typeface 9.5/15 Trump Mediaeval *System* QuarkXPress™ [SE]

A catalogue record for this book is available from the British Library

ISBN 0 521 80053 6 hardback.

Dedicated to the memory of

JAN VAN PARADIJS

Contents

Preface

'And what is the use of a book', thought Alice, 'without pictures or conversations?'

Gamma ray bursts suffer from an image problem. You cannot see them, you cannot feel them and nobody has ever heard them. An advertising agency would have a hard time selling this one. Even the name doesn't help. Let's face it; gamma ray bursts just aren't sexy. And one thing is sure; nobody is going to lose any sleep over them.

It really is a pity for the simple reason that, beyond any doubt, they are the most spectacular explosions in the entire universe, surpassed only by the big bang itself. They are the ultimate in cosmic fireworks, probably showing off their unseen grandeur every day somewhere in the universe. For anybody who gets their kicks out of extremes and superlatives, gamma ray bursts win hands down.

But for all that, cosmic super explosions do not conjure up images in the mind as do the big bang, when it ushered in the beginning of the universe, or the frightening force of black holes or the intriguing possibility of extraterrestrial life. There are only a few people who are really knowledgeable about such things and yet, in spite of it all, they are grist for the mill of science writers and have carved out a permanent place in our fantasy world. Write a book about one of these subjects and you don't have to tell anyone why it is interesting. But gamma ray bursts? All you get are raised eyebrows.

Flash! is the first popular science book devoted entirely to gamma ray bursts. And no wonder. The revolution that has taken place in the research of these giant explosions is only a few years old. In a remarkably short time the subject has metamorphosed from an obscure, apparently unsolvable puzzle into the hottest topic in astrophysics.

Every scientific revolution is accompanied by sweat, tears, and sometimes, even blood; and in the case of the gamma rays bursts it was no different. Ambition and jealousy, clique-forming and competition, triumph and tragedy – these are the ingredients that make for an adventure that has just begun, with no end in sight. These ingredients also make up the sum and substance of this book, a description not only of the science but also of the scientists who have contributed to the exploration of this astronomical phenomenon.

Flash! tells how this new and fascinating domain of astronomy came to be and at the same time offers a behind-the-scenes glimpse of modern astronomical research. Rather than being a boring world of stuffy professors, it is a dynamic and explosive work environment – the meeting place of observers, theoreticians and instrument builders – where the most powerful telescopes and the most sensitive detectors are used to unravel the mystery of the most cataclysmic explosions in the cosmos.

Science is people work. *Flash!* does not concern itself exclusively with gamma ray bursts but is very much about the people who have made it all happen. People of flesh and blood, with vision and with stories to tell. They are the lead players in a thrilling saga of discovery, a scientific adventure novel. Without their passion and perseverance gamma ray bursts would still be the puzzle they were thirty years ago. And without the enthusiastic cooperation of those people this book would never have seen the light of day.

I am particularly grateful to Carl Akerlof, Joshua Bloom, Sterling Colgate, Enrico Costa, George Djorgovski, Ed Fenimore, Marco Feroci, Jerry Fishman, Dale Frail, Andy Fruchter, Titus Galama, Neil Gehrels, Paul Groot, John Heise, Kevin Hurley, Ray Klebesadel, Chryssa Kouveliotou, Shri Kulkarni, Don Lamb, Andy MacFadyen, Chip Meegan, Hans Muller, Bohdan Paczyński, Jan van Paradijs, Hye-Sook Park, Luigi Piro, Martin Rees, Kailash Sahu, Brad Schaefer, Paul Vreeswijk, Ralph Wijers, Stan Woosley and Jean in 't Zand. *Flash!* is their story. Enjoy.

Acknowledgements

The illustrations were provided by: Carl Akerlof/University of Michigan (209 right, 222, plate 17), Joshua Bloom/California Institute of Technology (172), Jerry Bonnell/Compton Observatory Science Support Center (48), Hans Braun/Stichting Ruimte-Onderzoek Nederland (68, 73 left, 75, plate 4), Enrico Costa/Istituto de Astrofisica Spaziale (9, 63, 69), George Djorgovski/California Institute of Technology (119), European Southern Observatory (165, plate 13, plate 16), European Space Agency (253), Ed Fenimore/Los Alamos National Laboratory (8, 12), Gerald Fishman/Marshall Space Flight Center (10, 25, 27, 33, 137, 193, 202, 205, plate 1, plate 2, plate 3), Dale Frail/National Radio Astronomy Observatory (52, plate 11), Andrew Fruchter/Space Telescope Science Institute (122, plate 9, plate 18), Titus Galama/California Institute of Technology (99 right, plate 7), Peter Garnavich/Harvard-Smithsonian Center for Astrophysics (plate 15), Neil Gehrels/Goddard Space Flight Center (254, plate 20), Paul Groot (99 left), Astrid Havinga (140), Kevin Hurley (21), Instituto de Astrofisica de Canarias (95), Shrinivas Kulkarni/California Institute of Technology (145 right, 150), Donald Lamb/University of Chicago (41 right), Andrew MacFadyen/University of California, Santa Cruz (227 right, plate 19), Hans Muller/BeppoSAX Science Operations Centre (149, plate 8), National Aeronautics and Space Administration (plate 23), Bohdan Paczyński/Princeton University (41 left), Hye-Sook Park/Lawrence Livermore National Laboratory (209 left, 215, 218), Luigi Piro/BeppoSAX Science Data Centre (59 right, 80), George Ricker/Massachusetts Institute of Technology (plate 5, plate 21), Govert Schilling (145 left, 147, 179, plate 22), Piet Smolders/Artis Planetarium (plate 6), Space Telescope Science Institute (104, 134, 155, 232, 264, plate 12, plate 14), Stichting Astron (94), and Stan Woosley/University of California, Santa Cruz (227 left).

The extensive research for this book has been made possible by grants from the 'Stichting Fonds voor Bijzondere Journalistieke Projecten' in Amsterdam and the 'Stichting Wetenschap en Techniek Nederland' in Utrecht.

Prologue: The breath of Armageddon

'No, no! The adventures first' said the Gryphon in an impatient tone: 'explanations take such a dreadful time.'

Nature is cruel.

A star glitters in the full glory of its life. A clear, shining beacon of light and warmth, of life-giving energy. For millions of years this star lighted up the surrounding darkness, and planets basked in its gracious radiance. No one knows if life – the greatest wonder of the cosmos – has formed on one of these planets, but be that as it may, it is completely dependent upon this one star.

Then suddenly, it is over. Its time has run out. The star is in its death throes.

Armageddon is a Valhalla compared with what happens here. In a fraction of a second the star collapses in on itself. Trillions of tons of hot gas disappear forever through the one-way door of a black hole. Space becomes distorted, time ceases to have meaning, and matter is thrown into a frenzy. The star devours itself from the inside out. In a last agonizing scream the disappearing nucleus of the star spews out two jets of boiling, roiling matter as a desperate lifeline to the trusting world out of which it has been thrust.

With a speed of a billion kilometers an hour the two jets make their way to the outside in opposite directions. They bore burning tunnels through the outer layers of the star, which has no idea of the drama that has come to pass in its dark depths. But the star cannot continue to exist in the face of this cosmic violence. Like a Christmas bauble with a hand grenade inside, it explodes, spitting out an incredible quantity of energy that comes free from its inside.

The destruction is complete. The heavens rip open. A scorching, ball-shaped curtain of searing, glowing energy and expelled matter rushes with the speed of light to the outside. Planets evaporate like snow in the sun. Comets, moons, mountains and oceans, dead and living matter; everything is caught up in the ravaging fireball that shines more brilliantly than a hundred billion suns. What remains is a barren, abandoned battlefield, a desolate cosmic landscape of a dark void. It is the cosmic expression of the Last Judgment, the shadow of hell.

The cosmos does not trouble itself with the death of one star. Birth and death are the order of the day in the universe. Galaxies are in the process of forming; the big bang itself has just come into its own. The cosmos is still young and restless. Everything is in an uproar, balance has not yet found a firm footing. The explosion of the unfortunate star in a spiral arm of a far-removed galaxy is of no more importance than a fleeting tinseling of light upon the surface of a wave in the water, a single horror scene in the cosmic museum of curiosities.

Three billion years after the big bang the universe is still in its infancy. The Milky Way is just beginning to assume its form. There is not yet a trace of the sun and the earth. The present is still in the future, and the future stretches out over many billions of years. Billions of years during which countless stars will live and die, where matter is continually shaped and reshaped by the fundamental forces of nature, and where there will come life, intelligence, consciousness and a wonderful curiosity that will be capable of observing the very cosmos itself.

And even though the catastrophic death of the star quickly becomes history, its history is engraved in the legend of the universe far in the future, like a story that is told and retold from generation to generation or a book that is handed down through time. Within those few seconds that it took for the star to disappear from the stage, a gigantic rush of gamma radiation was blown into space. A burst of penetrating energy spread with the speed of light through the universe, like the deluge from a tidal wave that inundates the earth or the rumbling of thunder that rolls over the landscape.

The power of this gamma ray burst defies the imagination. In a few seconds more energy comes free than the sun emits in its entire lifetime. It is as if a million galaxies, each with a hundred billion shining stars, are packed together in a volume no larger than one million kilometers across. And as an expanding shell of radiation, the burst propels itself through the cosmos in all directions; it blows up like a balloon and travels with the speed of light three hundred thousand kilometers in one second, a billion kilometers in an hour, nine and a half trillion kilometers in a year. Although the exploded star is long forgotten and quiet is restored to the place of the calamity, the announcement of its death is delivered to every corner of the cosmos and the message of the disaster is carried far into the future.

Hundreds of millions of years pass by. In a remote corner of the cosmos a formless cloud of hydrogen gas slowly but surely collapses under its own weight. The cloud begins to rotate faster and even faster, until it flattens out, and in the fullness of time, the contours of a resplendent, spiral-shaped galaxy become visible. Our Milky Way is born – one of the infinite number of galaxies in the universe, lost among the billions of similar galaxies.

On its trip to the ends of the universe the photons from the gamma ray burst pass by untold numbers of galaxies. Some are large and stately like the Milky Way, others are small, unsightly misformed monsters, while still others resemble an illuminated rugby ball or a flat sparkling discus – galaxies in all shapes and sizes, bunched together in small, medium and large clusters. And in between there is a measureless void, vast, extended hollows of a dark vacuum, the deep black sea in the cosmic ocean separating those little islands of light.

When the storming wave front of gamma ray photons has transversed half the distance to our Milky Way galaxy, our sun gets born on the outer rim of one of the spiral arms, five billion years ago. It is a gigantic ball of glowing hydrogen gas just like all the other stars. Fired up by nuclear fusion reactions in its interior, it converts four million tons of matter into energy every second. But for all that, the sun is no more than a pin-prick of light compared with the giant star that met such an unfortunate end – nothing more than an unremarkable little dwarf star that would be invisible to the naked eye from a distance of just fifty light years away.

While the sun forms out of a rotating cloud of gas and matter, small cold cinders remain that circle the sun as cooled globes of gas or dark stony clumps, caught forever in the grip of gravity. On one of these cinders there are oceans of liquid water and drops of organic molecules that come down out of interplanetary space . There is now life on earth.

Whether or not life on earth is unique, we do not know. Or if the flash of gamma rays in its journey through space skims over other inhabited planets, no one can tell. But as the cosmic message comes relentlessly closer, the first multi-cellular organisms are in the oceans of the earth, the land is coloured green by the unremitting progression of the plant world and the amphibians creep up onto dry land. By the time the gamma ray photons reach the Local Supercluster – that extended swarm of galaxies of which our Milky Way is one – the reign of voracious dinosaurs holds sway upon the earth.

But the show is not over. There is another cosmic drama waiting to play itself out on this little planet of ours. A big ten-kilometer comet bores into the crust of the earth, playing havoc with the climate and doing the giant reptiles in, once and for all. Compared with the fireball caused by a gamma ray burst, this is a micro-catastrophe; a pebble hit by a grain of sand. But the aftermath upon the earth's biosphere is horrific: ninety per cent of all biological species die, and the evolution of life takes a new turn. The time of the mammals is upon us.

The racing gamma ray photons take no notice of all these happenings. Inexhaustibly they criss-cross the Local Supercluster, and with each passing second they come three hundred thousand kilometers closer to earth. When at last, in the far distance, our Milky Way galaxy becomes visible as a hazy smudge of light against the black starry sky, the earth's savannahs are peopled with the first predecessors of humans. But there are still a few hundred million years to go.

In the wink of a cosmic eye people learn to make tools, they discover fire, they develop voices and languages and begin for the first time to look in wonderment at the night sky and the stars above their heads that seem to exude eternal peace and perpetual rest. Primitive folk tell each other the myths of the constellations; Greek philosophers deliberate over the god-like perfection of the cosmos, and a Polish canon, who dares to question, contends that the earth is not the center of the universe. As the gamma flash passes the Pleiades, it is less than five hundred light years from the earth.

Now things really start moving. While the photons from the dead star race in the direction of the earth, Galileo Galilei looks for the first time at the star-laden firmament through a telescope, Isaac Newton drafts the law of gravity and William Herschel discovers that forms of light exist which cannot be observed by the human eye. The industrial revolution, the detection of the expanding universe, the first timorous steps on the way to space travel . . . As the gamma rays leave the bright star Capella behind, the first scientific satellites are launched and astronomers begin to study the cosmos through the use of invisible wavelengths.

And finally the speeding gamma ray flash zooms past Alpha Centauri, the neighboring star of the sun. The ten billion year trip of the gamma ray flash is almost at its end; there are less than four years to go before that signal sent out from the dying star reaches our living planet. In the meantime, artificial satellites are keeping an eye on the cosmos, and at the Kennedy Space Center in Florida preparations are underway for launching Vela 4, a military satellite equipped with gamma ray detectors.

On Sunday, the 2nd of July, 1967, a shower of gamma radiation blows through the solar system unseen and unheard, photons from billions of years ago and billions of light years away. Each one is a hundred thousand times as energetic as the visible photons of the sun. The greatest portion streams unhindered between the planets, and races on with the speed of light further into the dark heavens. A much smaller number enters the atmosphere of the earth, where they are absorbed by air molecules. And at last, a scant handful of gamma ray photons penetrate the detectors on the Vela satellite. The first detection of a cosmic gamma ray burst is now a fact. The mystery for man is just beginning.

I The sky watchers of Los Alamos

'You don't know much', said the Duchess; 'and that's a fact'.

Ray Klebesadel turned over another page. The numbers danced before his eyes. The pile of computer printouts, accordion-folded, was at least two inches thick. It looked like this was going to go on for some time. 'OK', he said, 'the next one I have is May 22nd, at 18.23.' Expectantly, he looked up from the paper at the man sitting across the table from him.

Roy Olson also had a big pile of computer printouts lying in front of him. He slid his finger down the rows of numbers. 'No, I don't have that one', he answered after a couple of seconds. 'The next one I have is May 25th at 04.17.' Now it was Klebesadel's turn. There was something on May 24th but nothing on the 25th. 'Nope; but maybe you have something on May 27th at 23.42?'

In their drab office at the Los Alamos National Laboratory in New Mexico it was still and stuffy. The date was March 1969; spring had just begun and nature was in full bloom. Outside the room, talking and laughter could be heard; a car door slammed shut. Don't get distracted. Keep going. Next line; the date, the time. Is it only on my list or do you have it too? How did we ever get started with all this?[1]

Ray and Roy worked through all the observational data from the two Vela 4 satellites. The analysis was done by hand; computers at that time were bulky and primitive. Line after line, page after page, searching for a phenomenon that both satellites had picked up at the same time. Ray had the readings from Vela 4A, Roy had the ones from Vela 4B. But up until now there was not one single event that showed up on both lists.

The Vela 4 satellites were launched on April 28th 1967 – two identi-

[1] Even though Ray Klebesadel and Roy Olson were indeed comparing the computer printouts in their Los Alamos office, the conversations related here are fictitious.

cal satellites, in high circular orbits around the earth at about an altitude of a hundred thousand kilometers. Each satellite weighed around 350 kilograms and was a good meter and a half in size. They were solidly packed with instruments for the observation of X-rays, gamma rays and electrically charged particles. Their most important job was to make sure the Russians weren't carrying out illegal nuclear tests in space.

1969 was the magical year of *flower power* and *love is peace,* but the Cold War was raging and the United States and the Soviet Union were locked in a feverish space race to show the other which one had the better technical know-how when it came to rockets. The space race, which at that time was a draw, was going to be won by America; within a couple of months Neil Armstrong and Edwin Aldrin would be the first men to set foot on the moon – perhaps the greatest technological achievement of the twentieth century. But the mood of the times said you couldn't trust those *commies*; who knows if, against all the international agreements, they wouldn't carry out nuclear tests on the far side of the moon, away from the view of the world. The Vela satellites were actually serving as nuclear policemen for the free West.

The Vela project dated back to the end of the 1950s. In 1958, the year after the world was shaken to its roots by the bleeps from the Russian Sputnik 1, for the first time within the United States there was talk of setting up a special commission to ensure the peaceful use of outer space: no war between the stars, no bombs on the moon,[2] and no nuclear tests inside or outside the atmosphere. But the politicians in Washington knew very little about bombs and nuclear weapons. They realized they had need of an expert and so they asked Sterling Colgate to serve as scientific advisor for the State Department.

Colgate, from the family of the famous toothpaste barons, was a physicist at Los Alamos National Laboratory where, during the Second World War, and cloaked in the deepest secrecy, the atom bomb had been developed. The laboratory was brought under the aegis of the

[2] In May 2000 it was revealed that the United States Air Force actually did have plans in 1958 to explode a nuclear bomb on the moon.

The Vela satellites were launched in pairs and separated from each other when they were in space. The many-sided satellites were covered with solar cells.

Atomic Energy Commission but was kept busy a good portion of the time with commissions from the Department of Defense. If ever there was a place where people were knowledgeable about explosives, it was at Los Alamos.

Colgate must have made a strange impression upon those government representatives. They saw a rather spare man, dressed in less than fashionable clothes, who said exactly what he meant and walked a bit gawkishly, like a teenager. But he knew how to point out the weak link in the UN proposition: if a ban were ever put on nuclear testing in space, there was no possible way to control such a treaty.

During the General Meeting of the United Nations in December 1959, held in Geneva, an official Committee on the Peaceful Uses of Outer Space (COPUOS) was established and the twenty-four members of the commission drafted the first version of what would later come to be the Nuclear Test Ban Treaty – an international treaty that forbade all nuclear tests outside the earth's atmosphere. And it was on the advice of Colgate that Los Alamos got the task of developing a satellite-borne sensor system to detect illegal nuclear explosions in space. 'Vela' would be the name of the system, coming from the Spanish word *velar*, which means 'to guard.'

Even though space travel was still in its early days, serious consideration was given to the possibility that the Russians would carry out their tests at the far side of the moon so that the Americans would be unable to detect them. The enormous blast of X-rays emitted in the explosion of a nuclear weapon would indeed be concealed by the moon. But the expanding cloud from the explosion would sooner or later have to be detected and the Los Alamos physicists knew that the fission products would continue to send out gamma radiation for some time. And so it was that Colgate stood behind the idea of outfitting the Vela satellites with gamma ray detectors.

But there was another problem. How would you know for sure if X-rays or gamma rays actually came from a nuclear explosion and not from the sun or another heavenly body? 'We'll look like fools if we don't

Ray Klebsadel (on the right) discovered the existence of gamma ray bursts in 1969 from observations made with the military satellite Vela. To his left are *Graziella Pizzichini* and *Chryssa Kouveliotou*.

know more', said Colgate to the Los Alamos advisor, Edward Teller. Teller, who once described nuclear weapon research as applied astrophysics, had no difficulty in rapidly developing an astrophysics program of its own at Los Alamos. Work was soon in progress based on Colgate's theory that supernovae – exploding stars – produce powerful bursts of gamma radiation.

Right about this time, in 1960, Ray Klebesadel, aged 28, came to work at Los Alamos as an electrical engineer. Klebesadel had studied physics at the University of Wisconsin at Madison and his professor, who had worked on the Manhattan Project, put in a good word for him with the Los Alamos staff. Klebesadel, who had only once in his life gone for a job interview, thought that the Vela project under the auspices of the United States Air Force offered an ideal work environment. Dedication, intellectual freedom and time pressure were the key words, and as a young scientist he could be involved in every phase of the project – designing, building, launching, data analysis, and anything else that came along.

The first two Vela satellites, referred to collectively as Vela 1, went into space on the 16th of October, 1963, the same year that the Nuclear Test Ban Treaty was ratified. Vela 2, which was also composed of two identical satellites, followed on the 17th of July, 1964. The first satellites had very simple detectors which were not particularly sensitive. Moreover, the observations were divided into segments of 32 seconds. If in this period (called the time resolution) six X-ray photons or gamma ray photons were registered, you didn't know if they arrived one at a time or all at the same time.

But the Vela instrumentation became more sophisticated very quickly. The Vela 3 satellites were launched on the 20th of July, 1965, equipped with more sensitive detectors and a time resolution of half a second instead of half a minute. A little less than two years later Vela 4 followed with a time resolution of an eighth of a second. The two Vela 4 satellites were not launched with the old Atlas Agena rocket but with the more powerful Titan IIIc rocket so that they could be more massive.

In March 1969 the Vela 4 satellites circled the earth for almost a year with Vela 5 standing on the launching platform at Kennedy Space Center ready for take-off. Klebesadel had just returned from Florida after the last pre-launch check. Finally, it was possible to place the data collected from the two Vela 4 satellites next to each other . . . '18th of June, at 08.09.' 'No. 20th of June at 19.36.' 'No, but there is one at 22.18. Do you have June 24th at 03.49?' Hang in, there are still a few hundred more pages to go.

There were more than enough reasons as to why only one of the two satellites registered in most cases. Fast-moving electrically charged particles also triggered a signal from the detectors, and even though the Vela satellites were outside the radiation belts of the earth, they were bombarded regularly by high-energy particles from the sun or from the universe. On the other hand, if the cause was a nuclear explosion in

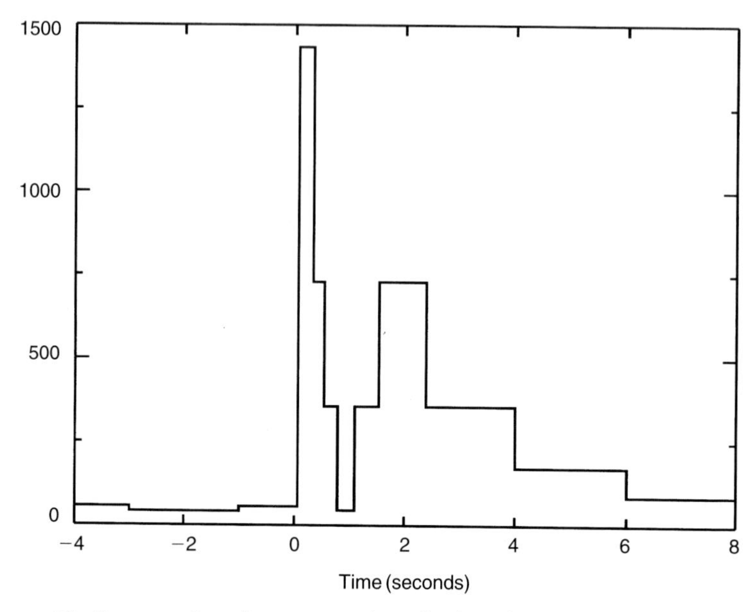

The first recording of a gamma ray burst by the Vela 4 satellite. The horizontal axis is the time, the vertical axis is the strength of signal. The burst showed a remarkable double peak.

space, an overwhelming solar flare or even an exploding star, its traces would be picked up by both satellites at the same time.

Again a new page full of numbers. 'July the 2nd, 1967, 14.19.' 'Yes! I have that one too! Precisely at the same time!'[3] Klebesadel and Olson looked at one another in astonishment. Then they placed the two computer printouts next to each other to compare the rows of numbers. At last they had a bite. No question about it: on Sunday the 2nd of July, 1967, the gamma ray detectors from both satellites picked up a strong signal; and it was a very exceptional signal, consisting of two peaks. The first one lasted less than an eighth of a second (the time resolution of Vela 4); the second peak lasted for two seconds – absolutely not what you would expect from the gamma radiation from a nuclear explosion.

Klebesadel grabbed a list to check old observation data from an

[3] Even though the conversations between Klebesadel and Olson are fictitious, the gamma ray burst on July 2, 1967, did take place at the recorded time (Universal Time).

American satellite that had registered solar flares. But on the 2nd of July there had been no solar eruptions. Nor did astronomers observe a new supernova in the sky on that day. Everything seemed to indicate that the Vela 4 detection was caused by an unknown source of cosmic gamma radiation. On closer observation, it turned out that the burst of gamma rays was also detected by the Vela 3 satellites, although the double peak was not too well defined because of the lower time resolution.

It was not surprising to find that there were sources of cosmic gamma radiation since it is actually similar to normal light, differing only in its extremely short wavelength and extremely high energy. Visible light to which our eyes are sensitive has a wavelength between 400 and 700 nanometers (one nanometer is a millionth of a millimeter). But in addition to visible light there are many kinds of electro-magnetic radiation that we cannot see, such as radio waves, microwaves, infrared radiation and ultraviolet radiation, X-rays and gamma rays.

Ultraviolet radiation is something we are all familiar with in our daily life. It is the ultraviolet (UV) radiation that makes our skin tan when we stay out in the sun. The pigment cells in the surface of the skin become activated by this UV radiation, something that does not happen in ordinary light. This tells us that UV radiation is more energetic than visible light. And if we need more proof, just remember that too much exposure to UV radiation can cause skin cancer.

Ultraviolet radiation has a shorter wavelength than visible light (between 100 and 400 nanometers) and the shorter the wavelength the higher the energy. X-rays have an even shorter wavelength (roughly between 0.01 and 10 nanometers) and are even more energetic. As anyone who has been to a dentist knows, 'soft' X-rays pass without a problem through human tissue but are somewhat restrained by teeth and bone. This is what makes it useful in medicine. 'Hard' X-rays are so energetic that living cells are impaired by them.

The most energetic radiation in all of nature is indeed gamma radiation, with wavelengths shorter than a hundredth of a nanometer (a hundred millionth of a millimeter). A gamma ray photon – a 'light particle' of gamma radiation – contains at least a few hundred thousand times more energy than a visible photon. Exposure to gamma radiation is what led to radiation sickness and death among the people of Hiroshima and Nagasaki in 1945.

Gamma radiation is rare on earth, although small quantities are emitted by radioactive materials such as uranium ore. The gamma ray photon is one of the by-products of the radioactive fission process where uranium atoms break apart to form lighter atomic nuclei. But in the universe enormous quantities of gamma radiation are produced by all sorts of explosive processes, such as solar flares and supernovae. Gamma radiation forms the rending, high tones of the cosmic symphony – shrill and penetrating.

Fortunately for us, the earth's atmosphere forms a natural shield against this deadly radiation from the universe. Gamma rays can fly right through the human body; they cannot be restrained even by a solid plate of lead. But the atmosphere forms an impenetrable barrier simply because it is tens of kilometers thick. In spite of the thinness of the atmosphere, a gamma ray photon on its way down comes up against so many atoms that a collision is unavoidable. The gamma ray photon then loses its energy and in fact gets absorbed by the atmosphere.

Therefore, cosmic gamma radiation can only be observed by using special detectors on board high-flying balloons or satellites, such as the Vela 4 scintillators. These are little blocks made of a special material – at first a kind of plastic was used but later they were made of cesium iodide. The atoms in these scintillators emit minuscule flashes of light when they are hit by a high-energy gamma ray photon. These tiny bursts of light are amplified and detected by photomultiplier tubes and the intensity of the flashes is directly proportional to the energy of the absorbed gamma photon.

Klebesadel and Olson were not quite sure about what they should do with their discovery. It was obvious that they were dealing with an unusual source of cosmic gamma rays but it wasn't possible to say where they came from, if for no other reason than that the scintillators on board the Vela 4 satellites were not direction-sensitive. And as far as the distance of the gamma ray source was concerned, they didn't have a clue. Although no solar flares were observed on July 2nd 1967, in principle it was possible that the gamma ray photons they saw came from the sun.

The Vela 5 satellites were now waiting to be launched. Klebesadel knew that they were more sensitive and had a better time resolution of one sixty-fourth of a second. If there really was talk of a new phenomenon then the decetectors of Vela 5 would observe many more gamma ray bursts. For the moment, it was decided not to tell anyone about the mysterious phenomenon but just wait for the results from the new satellites.

Vela 5 was launched on May 23rd, 1969, which happened to be the day of the 'roll-out' of the powerful Saturn V rocket that would lift off the Apollo 11 astronauts on their way to the moon. Nobody had time to give any thought to the relatively little Titan rocket with its military payload. Even Klebasdel found the roll-out more exciting.

But there is rarely instant gratification in space experiments. The results from Vela 5 were delayed. Through a fault in the electronics the detectors were actually so sensitive that they picked up an enormous amount of background signals. It was a hopeless task to analyze by hand all the signals that were registered from the mountains of measurements collected, as Klebesadel and Olson had been doing with Vela 4. Forced by necessity, Klebesadel learned to use the computer language of Fortran and wrote a program that would search through all the data to find the occurrence of *their* phenomenon on both satellites at the same time. Finally, they found about twelve that did not coincide with any solar flares or supernova explosions. Even more encouraging was the fact that some of the new detections showed the same remarkable double peaks that were seen with the Vela 4 flashes.

The last two Vela satellites were launched on April the 8th, 1970. Originally, the Vela program was to carry out five double launchings, but because it had been so successful it was decided to put the two back-up satellites into orbit. Vela 6 had the same instrumentation on board as Vela 5. The orbits were chosen to form a complete network around the earth by placing the satellites at the largest possible distances apart, in most cases many tens of thousands of kilometers apart.

Thanks to the great distance between the satellites, plus the high time resolution of Vela 5 and 6, Klebesadel was now able to determine the direction from which the puzzling gamma ray bursts came. Just like visible light, gamma rays travel through the universe at the speed of light: 300,000 kilometers per second. A gamma ray burst coming from a given direction would be detected just a bit earlier on one Vela satellite than on the other. The difference would be no more than a few tenths of a second, but because the arrival time was accurate to within a sixty-fourth of a second, it would certainly be observable.

Now that the arrival times could be carefully compared with each other, Klebesadel and his colleagues succeeded in tracing sixteen gamma ray bursts, regardless of where in the universe they had come from. It was immediately apparent that the flashes were not caused by the sun, the moon or the planets. The sixteen bursts were distributed completely randomly over the sky. It was also clear that they were from a great distance from the earth, beyond the solar system, which meant in turn that the explosions had to be monumentally powerful.

Together with Ian Strong and Roy Olson, Ray Klebesadel wrote an article about the discovery and on June 1, 1973, it appeared in *The Astrophysical Journal Letters*. A few days later, Klebesadel presented the impressive results to the 140th meeting of the American Astronomical Society (AAS) in Columbus, Ohio. For the first time the discovery of gamma ray bursts was made known to the world.

Thirty years ago the media paid less attention to scientific events than they do now, particularly when it had to do with such abstract topics as explosions of invisible gamma rays. Today, the AAS meetings are attended by any number of journalists and the more mysterious the subject, the greater the interest. But back in 1973 Klebesadel was approached by only one journalist, from the American tabloid, *The*

National Enquirer. Armed with the knowledge that the Vela satellites were intended to trace nuclear explosions from an enemy, the reporter asked if it were not possible that the gamma ray bursts came from a nuclear war between extraterrestrial civilizations. Devoid of any experience with the popular press, Klebesadel answered with typical scientific precision that the explosions did not in any way resemble nuclear explosions as we know them but in principle, he could not rule out the possibility. And so it was that the public-at-large became aware of gamma ray bursts. The screaming headlines could make anyone believe that *Star Wars* was a reality, even four years before the movie appeared.

The military character of the Vela program and the long period between the discovery of the first gamma ray flashes in the spring of 1969 and the publication in the summer of 1973 led to the rumor that the US Department of Defense kept the discovery secret and then waited four years before releasing the information. Even the Vela veteran Sterling Colgate thought for a long time that the satellite observations were the official property of the U.S. Air Force rather than the scientists at Los Alamos whose job, he thought, was only to analyze the data.

Klebesadel relegates the whole secrecy theory to the realm of fairy tales, even though he himself played a part in starting the myth. 'At the presentation of our results I showed a diagram that I made on the Vela project graph paper', he says. 'There were some codes on the paper which were totally innocuous, but the project leaders preferred that they not be made public. Upon the request of a superior I scratched them out with a penknife. But because the scratches were clearly visible, of course it was immediately believed that all sorts of secret information had been erased.'

At the same time, Klebesadel's presentation in 1973 caused a shock wave through the astronomical world. Tom Cline from NASA's Goddard Space Flight Center in Greenbelt, Maryland, was ready to tear his hair out because his detector aboard one of the IMP satellites (Interplanetary Monitoring Platforms) had picked up similar outbursts. Cline thought they came from supernova explosions and so he had not yet published anything about them. This was one time where there was no pay-off in erring on the side of caution. If he had been more assertive he would have gone down in history as the discoverer of gamma ray bursts. But he was not alone. Other astronomers also became aware that in their own satellite data there were all sorts of mysterious signals and unexplainable peaks.

How could the source of the gamma ray bursts really be determined? The answer was obvious. It was necessary to see if, at the place where the flash occurred, something very unusual would be visible – a remarkable star, for instance. Unfortunately, the positions of the gamma ray bursts in the sky were far from being accurately known. The sixteen 'localized' Vela flashes could, in the best of circumstances, be reduced to an area in the sky that was several times the apparent size of the full moon. An error box that size contains many thousands of completely ordinary stars. It is as if you were sitting in a packed concert hall and you heard somebody sneeze. If you looked in the direction the sound came from, you would only see a bevy of innocent faces.

To Kevin Hurley it was obvious that in order to determine the positions accurately, a network of space probes was necessary, placed at a distance of tens of millions of kilometers from each other. The arrival times at which a gamma ray burst registered on the various probes in the network would then differ by a number of seconds or even minutes, which would make a much more precise triangulation possible. Hurley had his first Inter-Planetary Network (IPN) ready to go and now, almost twenty-five years later, he is still imbued with the concept.

Hurley was born in New York but moved to California when he was ten. The young Kevin had from early on shown an interest in stories about the universe and so, one day his mother bought him an amateur telescope. When she took him to meet the famous astronomer Otto Struve he remembers that it made a lasting impression upon him. In the late 1960s, he went to study astronomy at the University of California at Berkeley where the Space Sciences Laboratory had just been set up. It was the golden era of space research; everything was new and money was no object. Hurley and other graduate students worked on a number of projects, one of them being balloon experiments to observe X-rays from the giant planet Jupiter. After he received his PhD in 1972 he went to Toulouse in France to work at the Centre d'Étude Spatiale des Rayonnements (CESR), where he helped develop the X-ray detectors for Russian scientific satellites.

The nine months he intended to stay there turned out to be fifteen years! Working in cooperation with the Russian Institute for Space Research in Moscow was pleasant and informal. Within a year and a half, he was free to think up a nice experiment, build the instrument and shoot the satellite off into space. It was when his director was attending a conference in Denver about cosmic rays that Hurley heard for the first time about the discovery of gamma ray bursts. It was immediately proposed that a special gamma ray detector be built for the Prognoz 6 satellite, which was launched on September 22,1977. This gamma ray detector was just the second type of instrument in the world that was especially intended to observe gamma ray bursts. It followed the one that Tom Cline placed on board the German space probe Helios 2, which was launched on the 15th of January, 1976.

It was relatively easy to put an instrument on board a Russian satellite, but to get the results quickly was a different story. The Prognoz satellites transmitted their measurements through a telemetry ship in the Black Sea. From there the tapes with the data went to the Institute for Space Research in Moscow, where they had to be copied in their entirety and then brought by diplomatic courier to the French embassy. And finally, via the embassy, the information came to Toulouse. When Hurley and his colleagues were in Moscow to examine the computer printouts, they would sometimes photograph them right in their hotel room because it saved so much time.

At the end of 1978 the first reliable Inter-Planetary Network became a fact. Five space probes were outfitted with gamma ray detectors: the Russian Prognoz 7, which had been orbiting the earth since October 30th, 1978, the German Helios 2, which was in an elliptical orbit around the sun, and finally, no less than three space probes used for studying Venus. NASA's Pioneer Venus Orbiter was launched on May 20th and had a detector on board that was developed by Klebesadel's team at the Los Alamos Laboratory. The Russian Venera 11 and 12 were launched on September 9th and 14th and were equipped with the detectors from Hurley's group in Toulouse. Along with all these there were also Russian Kosmos satellites circling the earth, some of which had a gamma ray detector that was built by the scientists at the Ioffe Institute in Leningrad under the guidance of Evgeny Mazets. Of course, the Vela 5 and Vela 6 satellites were still active.

By 1978 the gamma ray burst enigma was five years old, or if you asked Ray Klebesadel, actually nine years. It was high time that the mystery was solved and everybody expected that an accurate determination of the position of a gamma flash as revealed by the InterPlanetary Network (IPN) would lead to a break-through. Indeed, Hurley and his colleagues succeeded in determining the position of gamma ray bursts with an accuracy of a few arcminutes,[4] but even with the biggest telescopes in the world there was nothing of interest to be seen in those error boxes. To use the concert hall example again: it was as if you knew from which direction the sneeze came but strangely enough, nobody seemed to be sitting there.

On Monday, March 5, 1979, the tide seemed to turn. All the space probes from the IPN on that day registered an extremely powerful and long gamma ray burst. A few weeks later when the measurements were analyzed and the direction was confirmed, the burst turned out to coincide with a curiously shaped small nebula in the Large Magellanic Cloud, a neighboring galaxy of our own Milky Way at a distance of 160,000 light years (one light year is the distance light travels in a year at the speed of 300,000 kilometers a second: about 9. 5 trillion kilometers).

The nebula had been detected and cataloged much earlier and was known to astronomers as N49. It is a remnant of a supernova eruption:

[4] An arcminute is a 60th part of an arc-degree, and a degree is a 90th part of a right angle. An arcminute is further divided into 60 arcseconds.The apparent diameter of the full moon accounts for half a degree, or 30 arcminutes.

The Russian Venera space probe, to study the planet Venus, was part of the Inter-Planetary Network.

an expanding gas shell that is formed when a star explodes. Could there indeed be a relation between supernovae and gamma ray bursts? Hurley, Cline and Klebesadel's colleague Ed Fenimore were convinced that they were on the trail of the solution to the riddle. A supernova explosion could produce a compact neutron star (see Chapter 11), and in the strong gravitational field of such a neutron star all kinds of energetic processes could be set into motion which might produce gamma radiation.

But other astronomers, including Mazets and his colleagues, didn't believe it. According to them it was a coincidental occurrence that the two events happened at the same place in the sky. If the gamma ray burst had actually taken place in the Large Magellanic Cloud, then it had to be an incredibly huge explosion, otherwise a flash that came from a distance of 160,000 light years away could never be so striking. What kind of natural phenomenon could ever, in a couple of minutes, produce as much energy as the sun sends out in a thousand years? No, it seemed more likely that the gamma ray burst came from much closer by, at a distance not more than a few hundred light years, and it would be pure chance that it seemed to come from the direction of the Large Magellanic Cloud.

The conjectures about the March 5th flash lasted for a long time. Today, no one doubts that the burst came from the Large Magellanic Cloud. But it is now known that this was no 'ordinary' gamma ray burst (to the extent that you can use such a term for a phenomenon about which so little is known), but rather, it had to do with a so-called soft gamma repeater, an object that is discussed more fully in Chapter 12. The 'ordinary' flashes continued to be a mystery, as always, and the error boxes remained empty.

Ray Klebesadel is now retired and back in Green Bay, Wisconsin. He could never have imagined that 'his' discovery would give rise to a puzzle that would keep astronomers busy for thirty years. Once in a while he visits Los Alamos and now and then he goes to a conference about gamma ray bursts, but he is no longer actively involved. The fact that he may never see the mystery solved doesn't trouble him. 'I can die happily even if it isn't explained', he says with an affable laugh.

Kevin Hurley went back to Berkeley in 1987 and now has an office in the Space Sciences Laboratory with a magnificent view of the San Francisco Bay area. On a clear day you can see the Golden Gate Bridge and the Pacific Ocean. His database of scientific publications about gamma ray bursts, with which he got involved during the 1970s more as a hobby, now contains about 5200 articles but still, up to this moment, the mystery remains.

However, there is a very important difference now compared with twenty years ago. The number of observed gamma ray bursts is no longer a few dozen, but is now up there in the thousands, and the sky positions of a great many of them are known. These break-throughs have come in the 1990s, not with Hurley's Inter-Planetary Network, which incidentally is still active, but with two very special orbiting satellites and the continued enthusiasm and dedication of a number of researchers here on earth, some of whom were still waiting to be born when Ray and Roy were plowing through their computer printouts in Los Alamos.

2 The bat mystery

Alice laughed. 'There is no use trying', she said: 'one cannot believe impossible things.'

'I dare say you haven't had much practice', said the Queen. 'When I was your age, I always did it for half-an-hour a day. Why, sometimes I have believed as many as six impossible things before breakfast.'

There is something eerie about the Bat Cave, like the secret clubhouse in a boys' book. There are no windows and if the door were shut and the light went out, the place would be pitch black and stuffy. But fortunately the light is on and the door is open. Fourteen men and three women have crowded themselves into the little room; there are not enough chairs to go around. On the walls there are posters hanging up, along with newspaper clippings and photocopies of cartoons; on the table there are fat folders filled with indecipherable papers. But there is also some cake and candies because today is the birthday of one of the club members and in honor of the occasion he is wearing a shiny party hat.

Each morning at 11 o'clock there is a meeting in the Bat Cave. They are a motley group of people; everybody seems to be talking at the same time and cryptic terms such as 'spectral index', 'energy bins' and 'count rates' are flying around. Graphs with squiggly lines evoke cries of amazement each time a new column of numbers and letter codes is written on a whiteboard. In the corner there is a plastic celestial globe, one of the few indications that this 'boys' club' is indeed made up of a group of astronomers. At 11:40 the astronomical abracadabra stops. Eeveryone suddenly stands up and it is apparent that the meeting has come to an end. You almost expect somebody to lower the colors.

The bat is the mascot of BATSE, the most successful gamma ray detector of its time. Since its launching in the spring of 1991, BATSE (pronounced 'bat-see') has observed almost 3,000 gamma ray bursts.

(Left) Gerald Fishman was the principal researcher on the BATSE detector which was placed aboard the Compton Gamma Ray Observatory.

(Right) Charles Meegan.

Everything that is known to astronomers about the statistics of those mysterious explosions is due to BATSE and every day the new events that have triggered the detectors are discussed at the morning trigger meetings. How was the energy distributed? How many detectors picked up the sightings? Could it not have been a solar flare? OK then do we all agree that it really was a gamma flash? At this point the flash is given a number, the graphs disappear and a new folder is added to the ever growing pile of BATSE observations. Next trigger.

The Bat Cave is situated in the middle of the sprawling Space Science Laboratory of the NASA Marshall Space Flight Center in Huntsville, Alabama. In Huntsville you cannot avoid the presence of 'space travel'. In the arrival hall of the airport there hangs a huge scale model of the International Space Station that is in the process of being built. And from just about anywhere you look in the city you will see the 111 meter high Saturn V rocket that was placed as an eye-catcher in the summer of 1999 to attract tourists to the Space and Rocket Center.

It was here to the vast grounds of the U.S. Army Redstone Arsenal that the German rocket expert Wernher von Braun was brought during the 1950s to work on the rocket that would be the American answer to the Russian Sputnik. Von Braun's work was brought to fruition in January 1958 with the launch of the Jupiter rocket that placed Explorer I in orbit around the earth; the first successful American earth satellite. Later that same year, when the National Aeronautics and Space Administration (NASA) was founded, Von Braun was named the first director of the Marshall Space Flight Center, which was housed on the grounds of the military Arsenal. It is still engaged primarily in technological research on new rocket motors and propulsion techniques; science as such plays a lesser role at Marshall.

Gerald Fishman, who has been the principal investigator associated with the BATSE detector for the past twenty years, parks his car right by the main entrance to the Space Science building, a converted weapons depot. 'Whenever some reconstruction or renovation needs to be done, you regularly come across traces of mustard gas', he says as he pulls the heavy door closed. His Mercury Sable, complete with a licence plate bearing in big shiny letters the name BATSE, was left at home. Instead, today everyone admires his 1960 Chevrolet Impala. Fortunately, there is not too much traffic in Huntsville because the brakes on the Chevy are dangerously in need of brake fluid.

But Fishman is used to dealing with recalcitrant machines and apparatus. He learned his lesson well when he carried out his first balloon experiments. If you didn't have money for an expensive satellite and still wanted to observe radiation that could not pass through the earth's atmosphere you hung the instrument under a stratospheric balloon. To say the least, this is a risky business. You send up your apparatus, exposing it to the mercy of the elements, and after the flight, the gondola of the balloon drops down somewhere and you just hope that the most important parts of the instrumentation have remained undamaged.

When Fishman came to Huntsville at the beginning of 1973 he already had considerable experience with gamma ray detectors on

A balloon gondola with gamma ray detectors used in the early 1980s to observe gamma ray bursts.

board stratospheric balloons because of the time he had spent experimenting with them at Rice University in Houston. But it was at the Marshall Space Flight Center that the truly serious work began. Gamma photons pack a lot of energy but they are very scarce. If you want to study them in great numbers you need big detectors, not a scintillator of a few centimeters such as the ones on board the Vela satellites but one with a surface area of perhaps one square meter.

In June 1973, Fishman went to the summer meeting of the American Astronomical Society and heard Ray Klebesadel report on the discovery of gamma ray bursts. If they could see such an explosion each month with their small detectors, he thought, then we should be able to observe several each day with a bigger detector. Certainly, a bigger detector is more sensitive than a smaller one so it stands to reason that it should be possible to observe those flashes that are too faint to be detected by the Vela satellites.

Back in Huntsville, Fishman set up his own research group for cosmic gamma ray experiments and in 1975 the first 12-hour balloon flight was carried out with a detector that actually was the predecessor of BATSE. In spite of the considerably greater sensitivity of the scintillator not a single gamma ray burst was detected, nor did a second flight produce any results. In 1976, another member was added to the team – Charles Meegan, also from Rice University. Meegan was not a man who shied away from a challenge. You could always increase the size of the detector, and everything that scintillated could be mounted in the gondola of the balloon.

Finally, in 1980 and 1982, two long-lasting balloon flights were carried out, each with four detectors, with a diameter of half a meter, on board. The balloon that was sent up on May 29, 1982, stayed aloft for almost two full 24-hour periods and still nothing was observed (except solar flares). But Fishman and his colleagues did get a bite on Monday October 6th, 1980. During a 19-hour flight they saw a faint gamma ray burst, truly a paltry harvest when you consider that during the two balloon flights they had expected to see about forty.

Their expectations were based on the sensitivity of the detectors. Their big scintillators were about one hundred times more sensitive than the little ones on the Vela satellites. To put it another way, they could observe bursts that were one hundred times fainter than the faintest flashes observed by the Velas. In principle then, you could look ten times as far with the new detectors because a gamma ray burst moved to a distance ten times greater will appear one hundred times fainter. But to look ten times further means that you have a thousand times greater volume. So obviously you expect to see a thousand times as many gamma ray bursts. Since the Vela satellites registered about twelve gamma flashes per year, Fishman figured on twelve thousand per year which would mean about thirty every day. The two balloon flights to test this assumption lasted a total of 64 hours during which time they expected to see about 85 gamma ray bursts. And since, obviously, half of the flashes take place below the horizon then you come up with a possible score of something above forty. But instead of forty, Fishman and Meegan saw not more than one.

The reason, according to the researchers, was pretty obvious. If there is such an enormous shortage of faint gamma ray bursts then the greatest number of sources were apparently closer than the greatest distances they could cover. Imagine that you are standing one evening in the middle of a large meadow with chirping crickets all over the place. Anybody whose hearing wasn't too good would only hear the crickets nearby; the chirping of the other ones would be much too soft. However, someone with very sharp hearing could, in principle, hear the chirping up to a kilometer away, and you would expect that this would include an enormous number of soft cricket sounds. But if there are practically no soft chirps to be heard, this can only mean one thing: all the crickets are sitting close by, or in any case, certainly less than a kilometer away. Maybe the edge of the meadow – the boundary of the cricket distribution – is only five hundred meters away instead of a kilometer.

In 1982, no one had the slightest idea of the true nature of gamma ray bursts, but there was no dearth of theories floating around. Maybe they were caused by powerful explosions on neutron stars; these are extremely small, compact stars that radiate practically no light but have an enormous gravitational pull. Such neutron stars (in Chapter 11 we will come back to them in detail) are found all over the Milky Way galaxy; one could say they are the cosmic crickets in the meadow of the Milky Way. Whenever, with the help of a very sensitive gamma detector, you see far fewer faint flashes than you had expected, you can be pretty sure you are looking beyond the edge of the Milky Way galaxy. You have gone beyond the boundary of the gamma ray burst distribution. At greater distances in the empty space between the stars they are just not there.

If a sensitive detector is not able to observe more than one flash a day, then you are forced to come to the conclusion that balloon flights are not the way to go about studying this phenomenon. The ideal goal is to survey the cosmos continuously so that not a single flash is missed. Fishman's dream was to have a sort of Vela satellite, but with a detector area of a square meter. Within a year such an all-sky detector would be able to register several hundred gamma ray bursts and it would finally be possible to do a statistical study of the properties of these puzzling explosions.

In fact, at the end of the 1970s, Fishman and Meegan were deeply involved in making this dream a reality. NASA had plans to build a colossal gamma ray satellite that would be launched in 1985. Many interested research groups submitted proposals for placing an observational instrument aboard this Gamma Ray Observatory (GRO), and of course a sensitive gamma ray burst detector fitted quite naturally into the plan. In comparison with the other experiments, Fishman's proposal was rather modest: it would take up very little room; it would not use much electricity; it would be comparatively cheap; and, above all, it could be put together by a very small team. In actuality, it was nothing more than the usual balloon experiment that would simply be placed on board an earth-orbiting satellite.

The 'Transient Event Monitor', which is what Fishman first called the instrument, was selected in 1978 by NASA as one of the experiments to be included. But there were a number of things that had yet to be done. In order to scan the whole sky constantly you naturally need an array of detectors that must be placed on all sides of the satellite. Originally, Fishman wanted to build twelve detectors, each of which would be facing in a different direction. Six would be mounted on the top side of the satellite and six on the underside. NASA found this too expensive and preferred to keep the top side free for other, larger instruments. As far as Washington was concerned, six gamma ray burst detectors to be placed on the underside of the satellite were adequate, and just in the case of emergency, two more detectors would be built as backups.

But Fishman had his heart set on an 'all-sky' instrument. 'How about letting us place eight instead of six, one at each corner of the box-shaped satellite', he argued. NASA agreed, on the condition that he use the two backup detectors to make up the short-fall from six to eight. And thus was BATSE born: the Burst And Transient Source Experiment. 'I thought the word experiment sounded better than monitor', says Fishman about the name change; 'and furthermore, I wanted the word "burst" to be in the name because after all, that's what it's all about.' As far as any reference to bats is concerned, that came later from some Halloween toys. The bat became the mascot of BATSE and in the Bat Cave there hangs a bright yellow traffic sign with a black bat on it, carrying the warning 'Bat Crossing', to which someone added by hand, 'se'.

Eventually, a total of nine detectors were built; the backup one is now in one of the laboratory rooms of the Space Science building. It is quite large, about the size of a very big television set. The actual detector is a thick, circular scintillator, 51 cm in diameter and 1.27 cm thick. The scintillator is made out of sodium iodide, it weighs 100 kilograms and has a surface area of 2025 square centimeters. According to Fishman, it is no coincidence that his house address is 2025 and his telephone number ends with the same numbers.

Behind the scintillator, in a completely light-tight space, there are three large photomultiplier tubes which register the faint flashes that occur whenever the sodium iodide crystals are struck by an energetic gamma photon. But light flashes also occur when a rapidly moving electrically charged particle penetrates the detector – such cosmic ray particles are a thousand times more plentiful than gamma photons. To solve this problem a thin plastic scintillator was placed in front of the circular detector which reacts to cosmic radiation but not to gamma photons. If both scintillators give off a signal then it must be an energetic particle, but when only the main detector registers something, it must be a real gamma ray.

Even though BATSE turned out to be the star performer on board the Gamma Ray Observatory, in the beginning gamma ray burst research was looked upon more or less as second-rate science. By the end of the 1970s the flash mystery was almost 10 years old and there didn't seem to be a solution in sight. The novelty had worn off and a number of astronomers had lost their enthusiasm for the subject. Fishman had the feeling that each time he gave a presentation at NASA he was playing second fiddle. The other presenters, discussing the four large instruments to be placed aboard GRO, were allotted an hour while the BATSE team had to settle for half an hour.

But there were also advantages to being small and cheap. In 1982, when NASA was plagued by lack of room and budget over-runs, they had to let one of the experiments drop by the wayside. 'But we had nothing to fear', says Fishman, 'because dumping BATSE wouldn't make that much difference in space or money'. Finally, it was a sensitive spectrometer, 'GERSE' (Gamma-Ray Energetic Radiation Spectroscopy Experiment), that had to give way. This spectrometer would have very accurately measured the energy of each incoming gamma photon. In order not to lose this possibility completely, the eight BATSE detectors were later expanded and equipped with a small scintillator that could serve as a spectrometer. 'This is one of the few times in the history of all space research that an experiment, already in progress, was actually enlarged', says Fishman.

Meanwhile, the launch of the Gamma Ray Observatory, which was originally planned for 1985 was put off for about a year. Then again, in January 1986, the project suffered a severe delay when the tragic fatal explosion of the space shuttle Challenger brought the American shuttle program to an indefinite halt. For Fishman and his colleagues at the Marshall Space Flight Center the delay permitted them to make the BATSE design even better. Finally, in 1988 the eight detectors were delivered to NASA after an investment of 12 million dollars and the work of 400 person years.

Not that the Gamma Ray Observatory was launched in 1988 as was hoped. The delays in the shuttle program increased enormously and other scientific satellites got priority. Without a doubt, priority was given to the Hubble Space Telescope, the large optical telescope that was to be placed in orbit around the earth. The Hubble as well as the Gamma Ray Observatory were part of NASA's 'Great Observatories' program, which planned to have four giant space telescopes orbiting the earth: one to observe visible light (Hubble), one for gamma rays (GRO),

A rear view of one of the BATSE detectors.

one for X-rays and one for measuring infrared wavelengths.[1] Hubble was the first in the program and was launched on April 24, 1990, after which preparations began for the launching of the gamma satellite.

On April 5, 1991, it all came to pass. On board the space shuttle Atlantis the Gamma Ray Observatory left Kennedy Space Center and with the help of the shuttle robot arm, Linda Godwin, the mission specialist, lifted the monster out of the cargo bay of the shuttle to place the colossal satellite in its own orbit around the earth. Never before had such a heavy scientific satellite been launched. GRO is the size of a bus (9×4.2 meters) and weighs in at about 16 tons; this is in comparison with the Hubble Space Telescope, which is about 14 meters long and weighs only 9 tons.

[1] The X-ray satellite, the Chandra X-ray Observatory (CXO), was launched in the summer of 1999 and the infrared satellite, the Space Infra-Red Telescope Facility (SIRTF), is expected to be launched in 2002.

Shortly after its launching the GRO was baptized the Compton Gamma Ray Observatory (CGRO) in honor of the American physicist and Nobel prize winner, Arthur Holly Compton (1892-1962), a pioneer in the area of high-energy astrophysics. Despite the name change, the gamma satellite never achieved the popularity of the Hubble Telescope. The abstract graphics and contour plottings from the Compton observatory were no match for the spectacular color photographs of the universe being sent back from the Hubble.

But for many professional astronomers the observations from the CGRO were indeed of greater interest than the Hubble photos. Never before was the gamma ray sky observed in such detail. The four instruments together covered the entire gamma spectrum, where photons were detected with energies of 20,000 to 30 billion electron volts; in comparison, photons from visible light have an energy of not more than 2–3 electron volts. It was as if astronomers over the years had been studying a beautiful landscape from inside a bunker with small windows that only allowed them to look to the north, the east and the west using visible light, radio, infrared, ultraviolet and X-rays. When the Compton observatory was launched there suddenly came into existence an opening in the southern wall; a new window on the universe.

Within a few days of the launch all the instruments were calibrated and working. It was then that the BATSE Operations Room at the Marshall Space Flight Center began to register the first gamma ray bursts. Via a circuitous route the information came in within 24 hours of the event. The observations on the Compton observatory were first stored on tape recorders on board the satellite and a couple of times a day these were transmitted to large NASA communication satellites. These 'TDRSS' (Tracking and Data Relay Satellite System) earth-orbiting satellites sent the measurements through to a ground station on earth, and via another satellite connection everything finally ended up at the Goddard Space Flight Center in Greenbelt where the flight control center of CGRO is located. The very last step was to send the BATSE observations through to Huntsville. At first, they were sent via microwave link but later through the Internet.

Charles Meegan can well remember those first days. Almost every day a new gamma ray burst was observed. Some lasted no longer than a few hundredths of a second, others were minutes long. Some of the flashes had a rather smooth light curve: their brightness increased rapidly, reached a short-term peak and then gradually faded away. Other flashes flickered violently: the gamma ray intensity would vary greatly, even within a thousandth of a second. BATSE recorded all of this precisely because the electronics of the detectors on board were programmed in such a way that the brightness of a gamma ray burst could be measured more than one thousand times per second.

But there was something more important: BATSE could approximately determine the sky position of the unexpected explosions of gamma rays. 'As soon as a new gamma ray burst came in, the first thing we asked ourselves was what the galactic coordinates were', says Meegan. In other words, from which part of the Milky Way did the gamma ray burst come? By determining the positions of tens of hundreds of explosions, it would be possible to deduce some information on the distribution of gamma ray bursts in the universe.

The trick lies in the relative position and orientation of the eight BATSE detectors. They are each looking in a different direction and they have precisely the same orientation as the eight facets of an octahedron. Fishman wanted to get his team comfortable with this configuration so he had a number of big octahedrons made out of plastic and hung them up throughout the Space Science building. He also had special coffee mugs made for everybody, decorated with the same geometrical shape.

One gets an octahedron by setting two pyramids on top of each other at their base. Imagine that you have two small-scale models of the Great Pyramid of Cheops in Egypt each with a square as its ground and four triangular sides. Now in your mind, take these two pyramids and stick the two ground surfaces together. The result is a symmetrical eight-sided figure – an octahedron. Even though the eight BATSE modules are not actually stuck together but are mutually equidistant by a few meters, they have precisely the same orientation as the eight sides of the octahedron.

Anyone looking at an octahedron from different sides almost always notes that four surfaces can be seen from different angles. For the eight

BATSE modules the same thing holds true: anyone looking at the Compton observatory from different sides always sees four of the eight detectors. In other words, if there is a gamma ray burst anywhere in the universe, it will generally be registered on four of the eight modules. The number of photons that reaches any one detector depends upon the angle from which the detector is picking up the gamma rays. By comparing the strength of the signal in the four detectors, you can figure out from which direction the gamma rays are coming.

Actually, this method is not very precise, but in most cases the direction can be determined within a few degrees of accuracy. As each new BATSE flash was reported, its position was marked on a big map of the sky and within several weeks Fishman and his colleagues anticipated that they would be able to make some significant comment upon the distribution of gamma ray flashes in the universe. The expectation was that the faintest flashes, which most probably came from the greatest distance, would primarily be concentrated in that band around the sky which is our Milky Way.

This is similar to what happens with normal stars. All the stars that we see at night are part of the Milky Way galaxy – a big, flattened disc of several hundred billion stars. When one looks with the naked eye, only those stars near the sun are seen and they are visible in all directions. But if you look through binoculars or a telescope, stars very much further away come into view. These fainter stars are seen in a faint band that shows the true shape of the flat disc-shaped Milky Way galaxy of which the sun is a part.

If gamma ray bursts originate in the Milky Way galaxy, then they should appear to be distributed just like ordinary stars in the night sky. The brightest bursts, such as those observed by the Vela satellites, can in principle come from all directions, but the faint flashes must be concentrated in the plane of the Milky Way. If enough flashes were observed, the BATSE team expected that this 'Milky Way distribution' would become apparent.

But during the spring of 1991, slowly but surely it became clear that something remarkable was going on. The directions from which the bursts were coming were totally arbitrary; there was absolutely no pattern to be detected. Once in a while there was one nice flash seen in the Milky Way band but there certainly was no talk of the expected concentration. 'After about 12 bursts we were a bit surprised', says Meegan 'yet statistically speaking it didn't seem too significant. But after seeing that the 13th, the 14th and the 15th bursts also gave no indication of coming from the Milky Way, we began to get suspicious. Finally, after counting many more gamma ray bursts, we decided that something definitely was not right.'

Fishman had mixed feelings. 'On the one hand it was very exciting', he says, 'because it looked like we were on the track of an important break-through. On the other hand I worried about the consequences'. The puzzle of the cosmic gamma ray bursts was one of the biggest mysteries in astronomy but the BATSE observations, rather than clarifying the mystery, only made it more incomprehensible. The fact that there were far too few faint flashes observed (think of the crickets in the meadow) made one suspect that explosions were happening all over the Milky Way galaxy. But in that case one does not expect them to be arbitrarily distributed in the sky for the simple reason that the Milky Way galaxy has a strongly flattened structure.

The two most important statistical properties of the gamma ray bursts – the dearth of faint flashes and their distribution in the sky – simply contradicted each other. The first property shows that the instrument is sensitive enough to pick up the gamma ray bursts at the very edge of their distribution. The second property suggests that the gamma ray burst distribution is spherical and that we are looking directly from the center. 'There is no single population of objects in the Milky Way galaxy that displays this remarkable combination of properties', according to Meegan.

At first it seemed that there were only two possibilities left. Either the gamma ray bursts occur in the direct vicinity of the sun or they come from the depths of the universe. In the first case it could possibly be a phenomenon due to comets. The sun, at distances of many hundreds of billions of kilometers away, is surrounded by a tenuous cloud of hard frozen comet heads – ice balls not much larger than a few kilometers in size. This comet cloud, called the Oort Cloud is named after the Leiden astronomer Jan Oort who, in the middle of the twentieth century, established the existence of the cloud. If the gamma ray bursts occur in the Oort cloud we would see them on all sides around us. But because the Oort cloud is finite we would also see the edge of the burst distribution. Alas, there is no possible way known to create an energetic explosion of gamma rays on a deep frozen comet.

The second possibility is that the gamma ray bursts are not to be found in our own Milky Way galaxy but in other galaxies hundreds of millions or even billions of light years away. Far distant galaxies can be seen all around us and since we are unable to see endlessly far into the universe, it could be said that this in itself places a boundary on the distribution of gamma ray bursts.

In any case, it was very clear that an important discovery had been made with BATSE. This was also the opinion of NASA at its headquarters in Washington. On the morning of Monday, September 23, 1991, there was a workshop concerning the scientific aspects of the Compton Gamma Ray Observatory about to start in Annapolis, a town close to Washington. On this occasion NASA arranged a press conference in which the leaders of the various research groups could inform the public of the first results. 'The BATSE observations became big news,' said Meegan; 'somehow I can't remember the other presentations!' Fishman was no longer playing second fiddle but rather, he suddenly found himself to be the center of interest, particularly because he is a person who says what he thinks. The 'galactic' theory for gamma ray bursts, which claims that the explosions originate in neutron stars in the Milky Way galaxy, was in serious trouble as a result of the BATSE findings, said Fishman.

Neil Gehrels, the young, fresh project scientist of the Compton observatory, spoke out in the same vein: 'The neutron star model isn't dead yet but it's pretty badly in need of first aid', he told the reporters present. The next day all this made front-page news in the *New York Times* and the *Washington Post*, and if the court proceedings for the then potential Supreme Court judge, Clarence Thomas, concerning his suspected sexual harassment were not on the same day, Fishman would have been the studio guest on the popular television show *Good Morning America*. BATSE had made history by making an unfathomable puzzle even more of an enigma.

But it was not just the press and the public that were amazed and excited; Fishman's colleagues were practically struck dumb when, at the Annapolis meeting, he projected transparencies of 141 gamma ray bursts distributed all over the sky. Whatever the actual significance of the BATSE observations would turn out to be, of one thing there was no doubt: something revolutionary had happened in science.

The BATSE team then got busy writing a paper for the British scientific magazine *Nature*. It appeared in the January 9, 1992, issue and its cover showing a beautiful color photograph of the Compton Gamma Ray Observatory still hangs framed in Fishman's office. In reference to it, Meegan, the first author of the article, proudly says, 'One of the referees for the *Nature* paper said that ours was the most important one he had ever read and he felt it should be published immediately.'

In the stuffy Bat Cave there was much to discuss about the implications of the BATSE results. If indeed the gamma ray bursts came from distant galaxies billions of light years away from the earth then they must be extremely energetic, otherwise they could not be so easily observed. Fishman, Meegan and their colleagues could be on the track of cosmic super explosions, possibly unrivalled by any other explosion in the universe. Gamma ray bursts may well hold the key to the most energetic phenomena in the whole universe, indeed, to the most untamable primordial forces in nature.

3 A duel over distance

Tweedledee looked at his watch, and said 'Half past four'. 'Let's fight till six and then have dinner', said Tweedledum.

'Very well,' the other said, rather sadly: 'and she can watch us–only you'd better not come very close', he added: 'I generally hit everything I can see – when I get really excited.'

For Bohdan Paczyński the Annapolis workshop 'was the most exciting happening in my whole professional career'. For five years Paczyński, an astronomer at Princeton University, had been proclaiming that gamma ray bursts originated at cosmological distances in the universe – hundreds of millions of light years from the earth – but nobody took him seriously. Actually, it's no wonder, because his first ideas about the distance scale of gamma ray bursts were based on a mistake that he himself calls stupid. According to Paczyński, before the Annapolis workshop a good 99% of all astronomers were convinced that gamma ray flashes came from our Milky Way galaxy. Apparently, he constituted the remaining 1% because there were little more than a hundred astronomers involved in the subject at that time.

But he can remember that Monday afternoon in September 1991 as though it were yesterday. 'I sat there in the hall and I knew that within the next 15 minutes I would know whether I was right or wrong', he says. 'It was an incredibly tense moment – a question of winning or losing.' When Gerald Fishman showed that the distribution of gamma ray bursts was completely uniform and therefore they could not have originated in the Milky Way galaxy, Paczyński felt like he was the luckiest man on earth. 'For most of those present the BATSE results came as a big shock,' said Paczyński, 'but some colleagues came up to congratulate me.'

Paczyński, who wrote his first astronomy article while he was still in high school in Warsaw, became interested in the gamma ray burst

(*Left*) *Bohdan Paczyński* was one of the first who was convinced that gamma ray bursts came from far distant galaxies.

(*Right*) *Donald Lamb*, up until the beginning of 1997, firmly believed in the 'local' theory of gamma ray bursts, originating directly in the surroundings of our own Milky Way galaxy.

mystery in 1986. Every Tuesday at Princeton, there was an 'astronomy lunch' during which the most diverse topics were discussed. On one of those Tuesdays Paczyński's colleague John Bahcall was standing behind him in the cashiers' line in the cafeteria. 'I haven't got anybody to talk yet,' said Bahcall. 'Don't you have something nice to talk about?' Paczyński thought for a while and remembered a publication he had seen about gamma ray bursts. Five minutes later he was explaining to his colleagues that the available satellite observations did not match with the current theory, which contended that gamma ray bursts had something to do with neutron stars in the Milky Way galaxy.

And what did Paczyński base his contention on? After all, the Gamma Ray Observatory had not yet been launched and in 1986 there had only been about 100 bursts observed, mostly by the Vela satellites, the Pioneer Venus Obiter and the Russian Venera space probes. From none of these was the distance of the bursts known, so how could he say that they were cosmological and must come from far-distant galaxies?

It is a complicated story and as he thinks back now, Paczyński feels a bit embarrassed. 'It seemed to me that the old satellite observations already showed that there was a paucity of faint bursts,' he relates. 'That would mean that the gamma ray bursts originated in the direct vicinity of the sun or in distant galaxies.' Paczyński thought the BATSE discoveries in 1991 – a uniform distribution over the sky combined with a shortage of faint flashes – was already evident in the old satellite data.

As it later turned out, he had not interpreted the observations correctly. The satellite detectors reacted primarily to the peak luminosity of the gamma ray burst. A short burst with a high peak luminosity would of course register, whereas a longer, weaker one would not, in spite of the fact that the long burst perhaps produces in total more energy than the short one. This would make it appear as though there are fewer faint flashes, but in actuality it has to do with a selection effect of the detector.

'Bohdan has made brilliant contributions to astronomy,' declares Donald Lamb from the University of Chicago – an astrophysicist who, from the very beginning to the very last, believed in the neutron star theory – 'but he doesn't always pay enough attention to the observational data. When he came along with his cosmological theory, he was being laughed at by others behind his back. It didn't do his credibility any good.' Paczyński agrees. 'It may have been an understandable fault, but still, a fault. My one comfort is that I was not the first to be trapped. Two Russian researchers came to the same conclusion in 1975 but their publication in an obscure Russian journal didn't attract any attention.'

Nevertheless Paczyński continued to defend his cosmological model tooth and nail. In science it is a good habit to remain true to a theory and hammer away at it for as long as possible – it keeps your opponents on their toes and it serves to encourage scientific debate. But for Paczyński, unconsciously, something else played a role. Anybody who comes up with a deviant opinion attracts attention, and if it turns out to be correct, you win the jackpot. But on the other hand, if you go along with the crowd and it turns out to be right, you are just one among many and the best you get is a consolation prize.

Paczyński made a bet with Ed Fenimore who worked at the Los Alamos National Laboratory and had built a gamma ray detector for the Japanese Ginga satellite. He bet Fenimore a bottle of wine (quality not specified) that the gamma ray bursts did not come from the Milky Way but from far-away galaxies. Paczyński finally won the bottle but there was another bet he regrets not having taken on. In September 1991, right before the Annapolis workshop, he met the British theoretician Martin Rees at a conference in Germany. Rees contended that gamma ray bursts were a 'local' phenomenon. In other words, he believed that they originated in the Milky Way galaxy and at the time he was confident enough about it to bet Paczyński a hundred to one if it turned out to be otherwise. 'If I had only given him a hundred dollar bill,' sighs Paczyński now. Years later Rees said to him, 'We were both fools. I should never have made the offer and you never should have refused it.'

But those who think that people like Lamb and Fenimore were convinced by the BATSE results have it wrong. Yes, gamma ray bursts turn out to be distributed completely isotropically over the sky, and yes, BATSE 'sees' the edge of the burst distribution; but according to Paczyński's opponents, this doesn't necessarily mean that they can't come originally from neutron stars in the Milky Way galaxy. Indeed, at the end of the 1980s strong evidence was found to justify the 'local' model.

'In the beginning', says Lamb, 'I had my doubts. As a matter of fact, at a meeting in Aspen, Colorado around 1986, I showed an overhead of the Cheshire Cat from *Alice in Wonderland*. "The indications that gamma ray bursts come from neutron stars seems to be slowly but surely fading away like the smile on the face of the Cheshire Cat", I said. At the time I even advised my students not to get involved with gamma ray burst research – there wasn't much reward in it.' Lamb's colleagues were astonished that he, of all people, should hold such a pessimistic point of view.

A few years later the tide turned. Fenimore's gamma ray detector on board the Japanese X-ray satellite Ginga was able to measure the energy from the detected gamma photons rather accurately. And according to Fenimore some spectral lines were visible; at certain energies there were far fewer gamma photons seen than were expected. This was not the first time that astronomers thought they had observed the spectral lines in gamma ray bursts. Evgeny Mazet's Russian group had also reported such sightings in the early 1980s, but the Ginga observations were much more accurate.

Fenimore and Lamb declared that the spectral lines could only be explained by accepting that the gamma ray bursts originate on neutron stars with a very strong magnetic field. Neutron stars are the collapsed cores of exploded giant stars; and it has been known for a long time that they are not only small and compact (at least one and a half times as massive as the sun but not much bigger than about 20 kilometers across) but also have super-powerful magnetic fields.

Moreover, for some time, other satellites had also picked up many X-ray bursts in the Milky Way galaxy. It is true that X-rays are less energetic than gamma rays, but in both cases there have to be enormously powerful explosions. And from these X-ray bursts it was irrefutably shown that a neutron star with its tremendously strong gravitational field can cause gas from a companion star to land upon it at very high speed. A similar process might also be responsible for the more energetic gamma ray bursts.

Then too, it looked like some gamma ray bursts repeated themselves. The directions from which they came were not exactly known but a statistical analysis of all available data made one suspect that two bursts occurred at the same position in the sky more often than one would expect on the basis of chance. That also fit in very nicely with the neutron star model. If one explosion can occur on a neutron star, then why not more?

The idea that gamma ray bursts originate in distant galaxies would mean that there would have to be even more incredibly powerful explosions such as would be found, for example, in the complete destruction of a massive star or the merging of two neutron stars into a black hole. It would be out of the question that a second burst could be produced after a few months or a couple of years, for the simple reason that nothing would remain from the original body. Violent explosions are nothing new to astronomers but explosions capable of causing such total destruction – that goes a little too far. At a gamma ray burst meeting in Taos, New Mexico, in 1990, Lamb remarked jokingly that Paczyński probably regarded the name 'Taos' as an abbreviation for Total Annihilation Of Stars.

The BATSE results that were presented in 1991 in Annapolis came as a surprise to Lamb – he had expected that gamma ray bursts would be less uniformly distributed over the sky – but he wasn't bowled over by it. 'Those spectral lines from Ginga made the local theory very credible for me', he says. 'The slowly disappearing Cheshire Cat had been replaced by some pretty convincing evidence.' And many astronomers agreed with him. At the first Huntsville Gamma Ray Burst Symposium held in October 1991, the attending researchers were asked to vote and about half of them were of a mind that gamma ray bursts originated in the Milky Way galaxy, in spite of the remarkable results from Fishman and his BATSE team.

Not that there was much unanimity concerning the precise way in which the gamma ray bursts came to be. With the lack of detailed observations the theoreticians could give free reign to their fantasies and there were indeed, during the years, the most extraordinary models proposed: collapsing stars, colliding pulsars, 'earth' quakes on the surface of a neutron star, annihilation of anti-matter, the impacts of comets or asteroids on neutron stars, and many more, too phantasmagoric to contemplate. Robert Nemiroff from the George Mason University in Fairfax, Virginia, published in the 1970s a list of models of gamma ray bursts. He came up with a good hundred of them – considerably more than the number of bursts that had ever been detected up to that time!

At the beginning of the 1990s the situation wasn't much better. There were several hundred gamma ray bursts detected and the BATSE detectors were much more sensitive than anything that had been sent up into space before, but the mystery remained as mysterious as ever. Even after twenty years of research, it still wasn't known how far away they were, how much energy they produced nor where they came from. And in the meantime, new bursts were detected at an average rate of about one a day.

That there were so many theories floating around was seen by Paczyński not as the strength but as the weakness of this particular field of research. If you have a hundred theories and the right one is sitting in there someplace, he argued, then the other ninety-nine have to be wrong. And, he continued, if 99% of all the proposed theories are wrong, then the chances are that the last one also misses its mark.

Little by little the gamma flash world split into two camps, and the one wouldn't give the other the time of day. 'It even happened, that while I was giving a paper at a meeting about the neutron star model,' says Lamb, 'a colleague in the audience stood up, yelled "Total bull-shit", and stormed out of the hall.' Paczyński, too, was put down by many astronomers and was not taken seriously because they thought he didn't have a real feeling for the experimental world of satellites and detectors and paid far too little attention to the strong points in the physical models.

How to measure the distance of an unknown type of heavenly body has been a point of dispute since ancient times, and it is indeed one of the most difficult techniques in astronomy. In fact, the only distances that are accurately known are those of the celestial bodies in our own solar system and the stars in the neighborhood of the sun. Just a hundred years ago astronomers didn't even know if spiral nebulae such as the Andromeda nebula lay inside or outside our own Milky Way galaxy.

In 1994 Robert Nemiroff realized that there was a beautiful parallel between that old puzzle of the spiral nebulae and the present mystery of the gamma ray bursts. In both cases the astronomical community was divided into two camps and in both cases it was not known whether or not the objects were nearby or far away. The observations gave incomplete information or were amenable to two interpretations and emotions, then as now, often ran high.

Spiral-shaped nebula patches were known since the middle of the nineteenth century, and although many astronomers originally thought they were 'island universes', great collections of innumerable stars far removed from our own Milky Way galaxy, they began to think differently about them after 1870. Maybe the spiral nebulae were the birthplaces of new stars. Perhaps over time, out of the rotating cloud of gas, a star similar to our own sun could have been born.

In 1920 the conundrum was still unsolved; astronomers had heated discussions about the distance scale of the spiral nebulae and even about the size of our Milky Way galaxy. It was high time for an official scientific debate, thought George Ellery Hale, the astronomer who had made his name as the builder of the 2.5 meter telescope on Mount Wilson in California, the largest telescope in the world at that time. Hale approached George Abbott who was then secretary of the National Academy of Sciences, and after a bit of arm-twisting, he got Abbott to agree that a debate wasn't a bad idea.

On Monday the 26th of April, 1920, several hundred scientists came together – including Albert Einstein – in the Baird Auditorium of the stately Smithsonian Museum of Natural History on the Mall in Washington, D.C. Two well-known astronomers representing the two camps were each to give a 45 minute lecture explaining their theories. The 34-year-old Harlow Shapley from the Mt. Wilson Observatory defended the 'local' theory, i.e., that the Milky Way galaxy is very large and all spiral nebulae constitute a part of it, and the 47-year-old distinguished Heber Curtis from Lick Observatory told why he believed that the Andromeda nebula (and all other spiral nebulae) are actually galaxies like our own.

At that time the 'Great Debate on the Distance Scale of the Universe' attracted little attention from the press, but was later regarded by astronomers as one of the pivotal points in the history of cosmology. Why shouldn't such a 'Great Debate' about the distance

The most important participants in the Great Debate about the distance scale of gamma ray bursts: discussion leader *Martin Rees* (left), and the two opponents, *Bohdan Paczyński* (center) and *Don Lamb* (right).

scale of gamma ray bursts be held? Nemiroff believed that it should, and he organized, almost completely on his own, a 'Diamond Jubilee Debate' in the same auditorium and almost exactly on the 75th anniversary of the Shapley–Curtis debate. This was no small undertaking, particularly in view of the fact that it took place just a week before Nemiroff's wedding day.

Bohdan Paczyński and Don Lamb were the obvious candidates to debate the gamma ray burst theories. Martin Rees was asked to be chairman and moderator, and before the actual debate started, an introduction would be given by Gerald Fishman. Virginia Trimble of the University of California at Irvine would present an historical review of the original debate between Shapley and Curtis.

On Saturday afternoon, April 22, 1995, the Baird auditorium was packed to the hilt with 200 professional astronomers and interested others. Of course Ray Klebesadel, who had discovered the first gamma ray burst 26 years ago, was there, as was Willis Shapley, the grandson of

Harlow Shapley. No one believed that the answer to the distance of gamma ray bursts would be found on that afternoon, but nevertheless, there was a feeling of tension and anticipation in the air.

In 1920 Shapley and Curtis had simply given their two individual lectures – Shapley reading from his prepared paper, Curtis talking off the cuff and showing some slides of his text. But in 1995 there was more of a feeling of a real debate. Lamb and Paczyński would each give a 20-minute introduction, followed by a 10-minute period during which time each one could react to the other's standpoint and finally, another 5 minutes each to summarize. 'Shapley and Curtis were like two ships that passed in the night', says Lamb, 'while with us there was more tension and interaction, particularly because of the efforts and dedication of Nemiroff.'

In the end it was not just a clash between two points of view, but also between two cultures. Lamb represented the world of physics: the physical model of gamma ray bursts – explosions on the surface of neutron stars – were central to his argument. Paczyński was much more the classical astronomer. He concentrated totally on the few observable properties of the mysterious phenomenon, such as their distribution in the sky.

And of course there was the clash of personalities. Paczyński, the tall, energetic Pole with a cropped, astronaut-like haircut and intense, bright blue eyes behind metal-framed glasses, has an intellect that seems at times to be capricious and capable of taking agile leaps just as his body does. He has little patience with what he sees as stupidity among his colleagues. On the other hand there was Lamb, one of identical twins. He is a small, meticulously dressed, appealing man with friendly dark eyes; a man who signs off every e-mail with, *'with warm best wishes, Don'* and wants very much to please others.

How could Lamb ever equate his belief in the neutron star model with the results from BATSE? If the gamma ray bursts originate on the surface of neutron stars in our Milky Way galaxy they could not be distributed uniformly over the sky but rather, would clearly show a concentration in the plane of the Milky Way. Moreover, you would see more gamma ray bursts in the direction of the center of the Milky Way because the place of the greatest concentration of stars will also be the place where the most neutron stars are born.

Lamb's solution was the 'corona' of the Milky Way. According to him, just as the sun, during a total eclipse, is surrounded by a vast wreath of light, so could the Milky Way galaxy be surrounded by an enormously large, spherical 'cloud' of old neutron stars. So huge would the corona be, that the distance from the sun to the center of the Milky Way galaxy (about 30,000 light years) would no longer play an important role: even from our excentric position in the Milky Way galaxy, one would see an equal amount of neutron stars in all directions.

This idea differed considerably from the original neutron star model before the time of BATSE. Most pre-BATSE astronomers started with the idea that the brightest gamma ray bursts originated at a distance of a few hundred light years from the earth and therefore, they, just as the brightest stars, would be uniformly distributed throughout the sky. But the faint bursts also show a uniform distribution, in contrast to the faint stars, which are clearly concentrated within the bounds of the Milky Way. Therefore, Lamb had to accept that the bursts came from a far greater distance, possibly several hundred thousand light years away.

But of course, the question remained: how would neutron stars in the Milky Way galaxy ever get to be distributed in such a large, extended corona? They cannot have formed there because neutron stars are the remnants of exploded giant stars, and these you find only in the center and in the spiral arms of the Milky Way galaxy. However, in the 1970s astronomers had discovered that newly formed neutron stars sometimes have very high velocities and about a year before the debate, American radio astronomers even discovered one with a velocity of 800 kilometers per second, enough to escape from the gravitational influence of the Milky Way. Apparently, the stellar explosions from which they are formed are not completely symmetric and so, right at their birth, the compact, spinning stars are shot out and away at tremendous speed. With the passage of time, this is how a huge 'cloud' of escaping neutron stars around the Milky Way galaxy comes into existence.

Paczyński didn't believe in any part of this. There are billions of stars, galaxies and other objects in the universe, he told his audience, and hundreds of different types of heavenly bodies, but there is only a limited number of different sky distributions known to us. Objects in our own solar system – the sun, moon, planets and asteroids – show a clear concentration in one plane and are found almost all of the time in one of the twelve constellations of the Zodiac. Jupiter is never in the Great Bear and the moon is never in the Southern Cross for the simple reason that the solar system is flattened and we sit in the central plane. Likewise, all the objects in our Milky Way galaxy show a concentration towards the center and towards the central plane of the galaxy. You come across stars, star clusters and gaseous nebulae far more often within the confines of the Milky Way than outside of it, provided your telescope is sensitive enough to look into those far reaches.

There are only two isotropic distributions known, according to Paczyński: the brighter stars in the direct vicinity of the sun and the most distant galaxies in the universe. These are completely uniformly distributed over the sky. In view of the fact that gamma ray bursts also have such an isotropic distribution, they either have to be very nearby – no more than a couple of hundred light years – or very, very far away, at distances of many hundreds of millions of light years. And since it is highly improbable that gamma ray bursts exist only in the direct vicinity of the sun, the first possibility doesn't work. This leaves us with the cosmological solution – that the explosions occur in far away galaxies.

There was much to be said for both points of view, but Lamb had more arrows to his bow. If gamma ray bursts really originate in distant galaxies, those galaxies must be visible. With big telescopes galaxies can be seen at distances of hundreds of millions, or even several billions of light years. But the many searches for optical counterparts had turned up nothing. Via the Inter-Planetary Network set up by Kevin Hurley and his colleagues, the sky positions of gamma ray bursts were

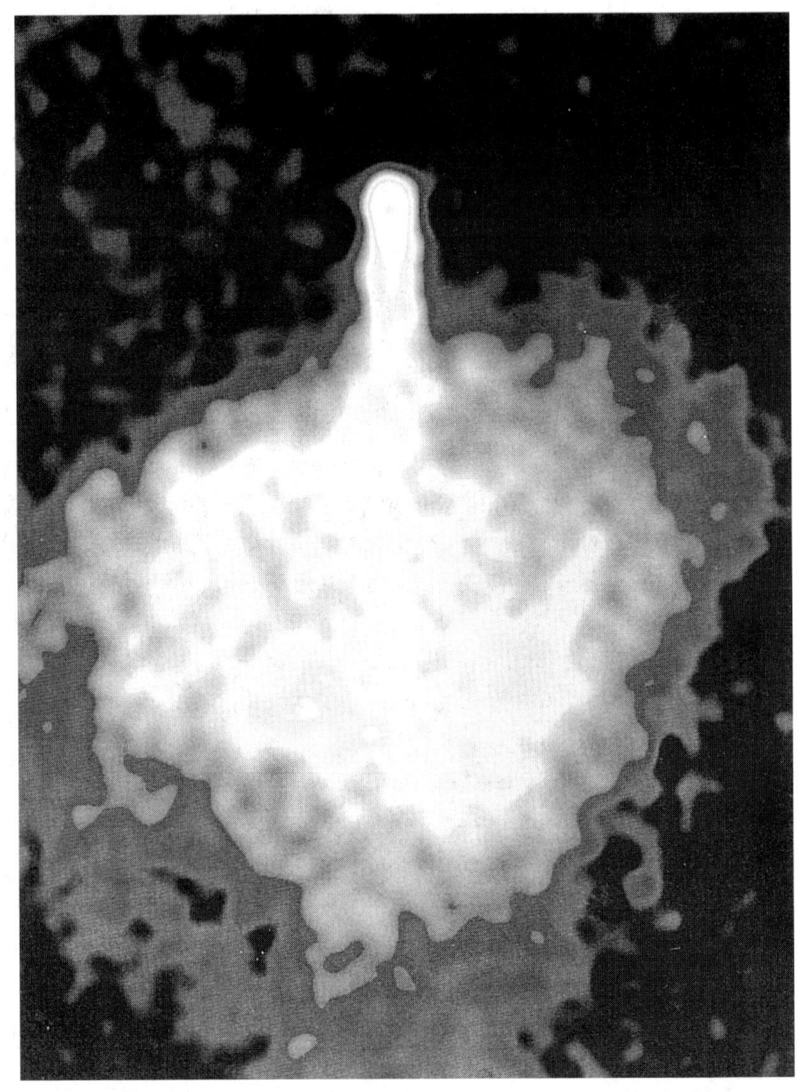

Radio observation of a neutron star moving at high speed through space. Thanks to their great speed, neutron stars can land up in the extended 'corona' of the Milky Way galaxy.

very accurately known, and yet, even with the most powerful telescopes in the world there was nothing in the small error boxes to be seen in the sky. But that would be exactly what you would expect to see if the bursts originated on the surface of neutron stars. A neutron star has a diameter of about 20 kilometers and is so small and dim that at a distance of only a few hundred light years it is no longer visible with an optical telescope.

Paczyński was not one to be caught easily. If our Milky Way galaxy has such a vast corona of old neutron stars, he reasoned, then other galaxies should also have them. You should expect that many gamma ray bursts come from the direction of the Andromeda nebula, a big spiral galaxy at a distance of 'only' 2.5 million light years. The Andromeda nebula looks very much like our Milky Way galaxy and the brighter gamma ray bursts in the corona of this cosmic neighbor should be visible to the sensitive BATSE detectors. But anybody who looks at the distribution of gamma ray bursts in the sky can see that there is absolutely no concentration in the direction of the Andromeda nebula.

Lamb knew very well how to parry the attacks from Paczyński. He even came up with a clever explanation to account for the 'missing' bursts from the Andromeda nebula. The energy of a gamma ray burst would primarily be sent out in two opposing beams coinciding with the rotational axis of a neutron star, he suggested. In order to see a flash, one of the beams must be more or less directed toward the earth. If you assume that the neutron stars at their time of birth are accelerated primarily along their rotational axis, either northward or southward, then the problem is solved. The neutron stars in the corona of our own Milky Way galaxy move away from us, so if they are far enough removed, their rotational axis always points more or less in the direction of the earth. If the gamma ray bursts are concentrated along the rotational axis then they are almost always visible from the earth. But in the corona of the Andromeda nebula there are only a few neutron stars in which the rotational axis points to the earth, by coincidence. Therefore, most of the gamma ray bursts in the Andromeda nebula cannot be seen.

'I had prepared very well for the debate,' says Lamb. 'Together with Ed Fenimore, a good friend since the beginning of the1980s, I spent a lot of time practicing. Ed played the role of Bohdan; I always had to have a retort ready.' Fenimore had complete faith in Lamb. 'We were much better informed than Paczyński about the weak points of the local model,' he says. And furthermore, what worked in Lamb's favor were the pretty overheads he had made and the humor he worked into his presentation. It wasn't long before he had the public in the palm of his hand.

At the end, Paczyński made a final attempt to bring the local model into discredit. 'During one of the intermissions I quickly made, by hand, an overhead to show that in the local model the gamma ray bursts move further and further away,' he relates. Around 1990 everyone assumed that the flashing neutron stars were no more than a few hundred light years away, in the flattened disc of the Milky Way galaxy. Now their hypothetical Milky Way corona forced them to relocate the neutron stars at a distance a thousand times greater. If the distance every few years must increase by a factor of a thousand, argued Paczyński, in a very short time there would be no difference between the local model and the cosmological model.

At the end of that memorable Saturday afternoon, it was difficult, on substantive points, to declare a winner in the debate. Nor was that the intention. But when it came to presentation and debating technique, Lamb was unquestionably the winner. Even dedicated protagonists of Paczyński were impressed by Lamb's arguments. Was it that many astronomers had a need to finally arrive at a credible explanation for the mystery? At least Lamb came with a well worked out physical model while Paczyński continued to contend that since the mechanism of the gamma ray bursts is unknown, the only thing you have to go on to determine their distance is their distribution in the sky.

'I don't know why the local model remained popular for so long', says Paczyński. 'Some astronomers had dedicated their whole scientific careers to it and that makes it difficult to be objective. But others were perhaps afraid of the consequences. If gamma ray bursts do indeed originate in distant galaxies, then the power of the explosions boggles the mind.' Super explosions that produce as much energy in a few seconds as the sun sends out in ten billion years seem to be too much of a good thing.

There was one point upon which Lamb and Paczyński agreed: the solution to the puzzle of the distance would come from new satellites with more sensitive detectors. More observations lead to better statistics. And if the bursts do indeed originate in the corona of the Milky Way galaxy then there must be a very slight asymmetry in the sky distribution because the sun and the earth are not in the center of the galaxy. It might even be possible that a sensitive detector would also discover gamma ray bursts in the corona of the Andromeda nebula.

There was also some hope placed in the possibility of finding counterparts at other wavelengths. If the sky position of the bursts could be determined quickly – faster than with the Inter-Planetary Network – then you could also search for them with X-ray telescopes, optical telescopes and radio telescopes. Who knows whether or not there is something extraordinary to be seen in the environs of the calamitous happening shortly after the explosion that would supply new clues and spur researchers off in another direction. As long as astronomers were only dealing with an enormous blast of gamma radiation without knowing precisely from which direction it came – and it only lasts a few seconds before disappearing – it would seem a hopeless task of ever tracking down the true nature of the bursts.

Three years after the debate between Harlow Shapley and Heber Curtis, the riddle of the distance scale of the spiral nebulae was solved. The famous American astronomer, Edwin Hubble (for whom the Hubble Space Telescope is named) discovered individual stars in the Andromeda nebula, which indicates that it must be a huge galaxy like our own. After the debate between Lamb and Paczyński, discussion leader Martin Rees said in his closing remarks that in this situation too, it was expected that the solution to the riddle of the distance scale of the gamma ray bursts would be found within a few years.

Although the BATSE detectors aboard the Compton Gamma Ray

Observatory had brought about a revolution in gamma ray burst research, they had not provided a convincing answer to the question concerning the actual nature of these phenomena. The hope of the American gamma ray burst community was now invested in a new NASA satellite, the High Energy Transient Explorer (HETE, pronounced 'Hetty'), which would be launched in the summer of 1994. HETE was much smaller than the Compton Observatory but it would be the first satellite that was totally devoted to gamma ray burst research.

That the solution to the mystery would finally come from an unknown European satellite which wasn't even designed to study gamma ray bursts was something nobody expected.

4 Liras, tears and satellites

It sounded an excellent plan, no doubt, and very neatly and simply arranged: the only difficulty was, that she had not the smallest idea how to set about it; and while she was peering about anxiously among the trees, a little sharp bark just over her head made her look up in a great hurry.

From the roof of the Telespazio building you can see the colossal dome of St Peter's far in the distance. But here, on the outskirts of Rome, it is tranquil; you hear the whisper of the wind and the song of the birds, and from the distance, the muffled noise of the traffic on the Via della Tiburtina comes wafting over you. However, it's not too hard to picture the chaos in the city. Tourists jostle against each other at the Trevi Fountain, modern pilgrims wend their way up the Via Conciliazione, guys and their girlfriends stroll along the Piazza Navona. Smoking, snorting buses stop on the parking lots near the Roman Forum, horn-blowing scooters recklessly maneuver their way through twisting, turning little streets and alleys, and embracing it all is the mysterious, imposing presence of centuries-old monuments like the Pantheon and the Colosseum. It is there that one finds the Piazza Campo dei Fiori, where Giordano Bruno was burned at the stake on February the 17th, 1600 – just a stone's throw away from the Vatican – because he was convinced that the stars in the night sky were far-removed suns, each with its own planetary system.

If the eyes of Pope Clement VIII, who was Pope at the time of Giordano Bruno, were sensitive to cosmic gamma rays, the daily bursts in the heavens would have told him that the firmament is far less eternal and unalterable than the Church was preaching at that time. But in spite of the overwhelming power of gamma ray explosions in the universe, they could not blast their way through to the Dome of St Peter's; not then and not now. They can only be seen on the computer monitors at the BeppoSAX Science Operations Center on the first floor of the Telespazio building when the X-ray cameras of BeppoSAX happen to be facing in the right direction.

That there would be an X-ray satellite capable of detecting gamma ray bursts was something nobody in the United States could have anticipated. BeppoSAX – wasn't that the impossible Italian project that kept taking longer and costing more money? No one believed that the satellite would ever get off the ground, let alone get into space. In books about the future of X-ray astronomy, you would never read anything about BeppoSAX; publications discussing gamma ray bursts never mentioned that particular satellite.

Enrico Costa, from the Italian IAS (Istituto di Astrofisica Spaziale) can laugh about it today. Anybody who has ever come to Rome knows that the Italians can't even organize a good shuttle bus to the airport, he says, let alone build a scientific satellite. Giovanni Bignami, director of science for the Italian space agency ASI (Agenzia Spaziale Italiana) once exclaimed in desperation: 'Who figures out the correct orbit? Who will be in charge of the telemetry?' ASI had its doubts. Italy had no experience and not too much understanding of these things.

But after a ten-year delay and a budget overrun of 100%, believe it or not, it succeeded. On April 30th, 1996, BeppoSAX was launched, and in less than one year, the mystery of the distance scale of gamma ray bursts was solved, thanks in part to Enrico Costa's Gamma Ray Burst Monitor.

Costa lives in a quiet quarter in the northeast of Rome, a few kilometers from the Porta Pia in the old wall of the city. With his dark eyebrows, gray beard and mustache, lined face and thick seaman's sweater, he looks more like an old fashioned sea captain than a scientist. He mutters with a heavy Italian accent and when he tells you something, not only are his eyes piercing and his big hands gesturing effusively, his whole body actually gyrates to add expressive emphasis to his words.

For Costa it all began in 1980 when he was sitting with Livio Scarsi, stuck in Rome traffic. Scarsi, a physics professor at the University of

(*Left*) *Enrico Costa*, principal researcher on the Gamma Ray Burst Monitor on board the BeppoSAX satellite.

(*Right*) *Luigi Piro*, project scientist on the BeppoSAX satellite.

Rome, 'La Sapienza', was busy setting up a research group for high-energy astrophysics, just as Gerald Fishman had done a few years earlier in Huntsville. Scarsi and his colleagues had experience with balloon experiments but when he told Costa about his plans to build their own Italian satellite that really came as a surprise.

The *Satellite per Astronomia X* (SAX, satellite for X-ray astronomy) was to be launched in 1986 and would cost about 28 million US dollars. The most important instrument on board would be a scintillator that was sensitive to energetic X-rays. It would not be made from plastic or sodium iodide like Fishman's scintillators; it would be a high pressure tank filled with a mixture of the noble gases xenon and helium, which would produce fluorescence as soon as an energetic X-ray photon interacted with it.

Even though there were doubts from the very beginning regarding the feasibility of the project, SAX became more and more ambitious. In addition to the gas scintillator there also had to be spectrometers and telescopes for low and medium X-ray energies, and, most desirable, a detector for extremely hard X-rays at the boundary area of gamma rays. SAX would be the first satellite in the world that could observe the whole X-ray spectrum from ten to one hundredth (0.01) nanometers.

It is no wonder that it took years for SAX to get further than the drawing board. It was quickly realized that the planned launch date would never be met and after the explosion of the Challenger in January 1986, it was decided to totally redesign the satellite so that it could be launched with a regular booster rocket. Meanwhile, the costs were soaring out of sight. It was time to look for collaboration with the Dutch space agency, the Netherlands Agency for Aerospace Programs (NIVR). The Dutch were very experienced in building X-ray cameras, having done so for the Russian space station, Mir, and the Dutch Wide-Field Cameras would give SAX a much greater field of view than the Italian instruments.

Eventually, it would be these Wide Field Cameras developed and built by SRON, the Space Research Organization Netherlands in Utrecht, that would transform BeppoSAX into a gamma ray burst observatory of the first order. But without Costa's Gamma Ray Burst Monitor (GRBM) this would never have been possible. The GRBM registers a gamma ray burst and can roughly determine from which direction it comes. If the particular part of the sky happens to be in the field of view of one of the two Wide Field Cameras, the X-ray observations are scanned for a simultaneous X-ray burst. If one is found, then its position in the sky can be much more accurately known. The satellite is turned in the right direction and the fading burst can be studied in detail with the other X-ray instruments on board.

In the original plans SAX was not going to have a gamma ray detector at all – after all,you can't study the entire electromagnetic spectrum with one satellite, and trying to cover all possible X-ray energies was ambitious enough. But the detector for extremely hard X-rays (the Phoswich Detector System) had to be shielded from photons and electrically charged particles which would be forced in from the side, and Costa realized that the four 'protective shields' could serve beautifully as gamma ray burst detectors. The ten millimeter thick shields are made of cesium iodide crystals that also react to gamma photons.

Costa brought his proposal to Fillippo Frontera, the leader of the Phoswich team, who was immediately enthusiastic about it. Indeed, it would be a shame if SAX didn't carry out gamma ray burst research, even though the satellite was not actually designed for it. Maybe it would finally be possible, right after a burst, to get an accurate fix on its position in the sky so that astronomers with big telescopes on earth would not have to wait a couple of weeks until Kevin Hurley's Inter-Planetary Network came up with a position.

And so SAX slowly but surely evolved from a modest little national satellite with one gas scintillator into an impressive Italian–Dutch space observatory that, with its Wide Field Cameras and telelenses, should be able to scan the skies at all possible X-ray wavelengths and moreover, would have the potential to solve the greatest mystery in high-energy astrophysics.

In the meantime, United States astronomers were not sitting idly by. The American gamma ray burst scientists were completely aware that accurate sky positions were the key to the puzzle. The Compton Gamma Ray Observatory would not deliver those accurate positions; the directional sensitivity of the BATSE detectors was by far not good enough for that. In spite of the intense lobbying of NASA by astronomers to put an X-ray telescope aboard CGRO, they were unsuccessful. Compton was no X-ray satellite, it was a gamma ray satellite and that was how it was going to remain, said NASA.

Thus it happened that in 1983 a group of astronomers put their heads together and plans were forged to develop their own gamma ray burst satellite. The fact that in far away Italy there were people busy with the same idea was not known to anyone. Stan Woosley, a supernova expert from the University of California at Santa Cruz, organized a workshop in July of 1983 on cosmic gamma ray bursts and X-ray explosions. During the meeting Woosley talked with Don Lamb from the University of Chicago and Ed Fenimore from the Los Alamos National

Laboratory about the possibilities of a High Energy Transient Explorer (HETE).

HETE started out as an ambitious plan but dwindled down to a small, inexpensive satellite with four gamma ray detectors, an X-ray camera and four electronic cameras for visible light and ultraviolet radiation. Naturally, the satellite would not be able to cover the whole sky at the same time but thanks to the reasonably large field of view of the instruments it would be possible to get a gamma ray burst in view once every few weeks. If, during the burst, X-rays, ultraviolet radiation or visible light was emitted, then the HETE cameras would probably be able to determine the sky position and pass it on to the observers on earth.

But HETE also remained a drawing board satellite for a couple of years. Most of the scientists involved in the project had little or no experience in presenting such a comprehensive proposal to NASA. Kevin Hurley from Berkeley, who was later to be very much a part of the project, had worked together with the Russians; Fenimore from Los Alamos had worked with the Japanese and of course he had experience with the Vela satellites for the Department of Defense, but getting support for a civil U.S. space research project was another kettle of fish. Fortunately, George Ricker from MIT (Massachusetts Institute of Technology) in Cambridge, Massachusetts, had dealt with such problems before. Ricker offered to become a member of the HETE team and in 1985 he succeeded Woosley as principal investigator of the project.

It was not until 1986 that an official proposal was finally submitted to NASA. After the satellite program was completely reviewed in detail, an affiliation was sought with institutes in France and Japan. In contrast to the Italian–Dutch Satellite per Astronomia X, which had, in fact, grown into an all-round X-ray satellite, the High Energy Transient Explorer was brought back to being a satellite that could do only one thing well: localize the direction from which gamma ray bursts had come.

In 1990, one year before the launching of the Compton Gamma Ray

Observatory, NASA finally allotted the money for HETE and work could really begin on building the new satellite. The project was to cost 14.5 million dollars, and not one cent more. If anything untoward happened, then the team would have to start eating 'dry bread and water', as Lamb put it. The launch was planned for the summer of 1994.

While the Americans got busy building HETE the Italians too were plowing ahead with SAX. But while HETE had a fixed price cap, the costs of SAX were soaring. The satellite itself finally cost 200 million US dollars; the whole project, including launch and ground station costs amounted to something like 350 million US dollars. Where would ASI get that kind of money?

Not to worry. In a land where the mafia is perhaps more powerful than the Pope and where cabinet crises are about as frequent as gamma ray bursts, it is also possible to come up with marvelously creative ways of getting around money and the law. Even though ASI, according to its own legal statutes, was required to spend 15% of the yearly budget allotted by the government on scientific research, Luciano Guerriero, chairman of the board of directors, put a part of that 15% toward the financing of SAX, primarily to keep in the good graces of the national space industry. SAX was a very attractive commission for Alenia Spazio and the satellite builders in Rome would not be too thrilled to see the project go down the drain. But to fulfill the letter of the law on the 15% ruling, Guerriero booked the yearly contribution to the European Space Agency (ESA) amounting to the sum of 385 million dollars, under the heading of research.

In 1991 this wheeling and dealing led to an enormous row between Guerriero and Remo Ruffini, the chairman of the scientific commission of ASI. Guerriero and his board of directors were accused of mismanagement, corruption and financial misconduct, and SAX was seen as the reason for all the misery. The controversy completely incapacitated the Italian space agency. To add insult to injury, a commission of 'wise men', which had been set up by the then minister of research, Sandro Fontana, came to the conclusion, within a month, that

Guerriero was not to blame. But Ruffini and his group could not let this rest, and after a long judicial procedure the top management of ASI, in 1993, was officially charged. One year later, the same fate befell Fontana himself. The Italian Treasury declared that the minister should never have been satisfied with the unseemly haste and irresponsible methods of the research commission.

Even though justice triumphed in the end and ASI, in the beginning of 1994, started up again with a totally new board of directors, the relationship between SAX and the rest of the Italian scientific world was ruffled. The satellite was seen as a prestige project that had got out of hand and, moreover, the enormous delay would make it out of date by the time it was ready to be launched. According to top scientists such as Nobel prize winner Carlo Rubbia and the celebrated X-ray astronomer Riccardo Giacconi, there would be no role left for SAX, what with the launch of the German Rosat back in 1990 and the Japanese ASCA (Advanced Satellite for Cosmology and Astrophysics) in 1993.

For Enrico Costa and the other SAX researchers these were frustrating times. 'You have to be a sort of Mohammed Ali,' says Costa, 'be able to stand on your own feet until you can deliver the lethal blow to your opponent.' Wonder of wonders, SAX finally came through the battle unscathed and in 1995 it looked like the launch could take place the following year.

When the actual launch date came into view it seemed logical to change the satellite's name. Wouldn't it be nice to name it for a famous Italian X-ray astronomer? No, of course it couldn't be Giacconi – and besides, you never name satellites for people who are still living. But it could be named for Bruno Rossi who, with his colleagues, discovered the first cosmic source of X-rays. Rossi had been a professor of physics at MIT and was recognized all over the world as a pioneer in X-ray astronomy and cosmic plasma physics. He died in 1993.

But there were more people who had come upon the same idea. On December 30th, 1995, NASA launched an X-ray satellite that could observe very rapid variations in the luminosity of cosmic X-ray sources, and in February 1996 this satellite was officially christened the Rossi X-Ray Timing Explorer (RXTE). But it was no problem; Italy had produced more renowned physicists. The choice of the SAX team went quickly to Giuseppe (Beppo) Occhialini, a contemporary of Rossi's, who also died in 1993. Occhialini was one of the founders of cosmic ray research and gamma ray astronomy. Moreover, he had played a major role in creating the European Space Research Organization (ESRO), the forerunner of ESA.

Looking back on the choice of name, Costa now finds it a good one. SAX was of course intended to be an X-ray satellite so in the beginning he regretted that the name of Rossi had been claimed by the Americans. But now that the Italian satellite has become world famous through its research on gamma ray bursts, he feels the name of BeppoSAX was absolutely appropriate.

In February 1996 Costa went to the Kennedy Space Center in Florida where the last calibrations and tests were being carried out before BeppoSAX was launched on April 30th. While down there he got the scare of his life. It seemed that the Gamma Ray Burst Monitor (GRBM) registered an enormous number of X-ray photons – around 5000 per second! In such a 'noisy' environment the signal from a cosmic gamma ray burst from a great distance in the universe could never be observed.

There was no doubt about what was causing it. In the satellite there are several well-calibrated radioactive sources that are necessary to gauge the measuring apparatus. They send out a specifically known quantity of X-rays and in this way it is possible to know precisely how the different detectors are responding. But the calibration source of the gas scintillator somehow 'leaked' X-rays and 'blinded' the GRBM.

Costa immediately contacted the project management at ASI. Either another calibration source had to be put into place or they had to find another way of shielding the GRBM. Unless radical measures were taken, the whole gamma ray burst venture of BeppoSAX could go under. But that was not what ASI wanted to hear. So close to launch time was not a time to experiment, they argued. Permission was not granted to tinker around with the satellite.

'ASI had the idea that the satellite was kept in a glass box for the two

BeppoSAX in the test room at the Kennedy Space Center in Florida.

months before the launch,' complained Costa, 'while in fact, everybody is busy with screwdrivers and spanners.' The technicians from Alenia Spazio who were in Florida didn't give up so easily. We may not be allowed do anything, they reasoned, but nobody can stop us from looking into the problem. 'We absolutely immersed ourselves in the records to get all the information we could about the radioactive sources,' says Costa. 'It was an enormous job when you realize that for every kilogram of hardware sent up into space there is about a ton of paper produced on earth.'

The search produced nothing – there was no information to be found. What now? There was only one solution left: in one of the 'clean rooms' there was a spare gas scintillator with its own calibration source and actually, according to the letter of the law, ASI hadn't forbidden them to tinker with this backup. 'When we took out the source, we knew at once what the trouble was,' says Costa. 'The cover plate was made out of aluminum instead of lead.' The problem on board the satellite could be cured with a very simple alteration. Costa got permission from ASI and within 24 hours the X-ray leak was sealed.

On Tuesday April 30th, 1996 – Queen's Day in the Netherlands – the Italian–Dutch BeppoSAX was lifted by an Atlas G Centaur rocket into a 600-kilometer-high orbit around the earth. 'Up until 15 minutes before the launch,' says Costa,'I still couldn't believe that it was actually happening. For more than 15 years I had learned to live with the thought that SAX may never fly. But here we are; it has come to pass.'

Italy's 'impossible' satellite became a reality against all expectations. Five years later, flight managers and scientists are still in daily contact with BeppoSAX, which, once every100 minutes, orbits the earth, almost exactly above the equator. The so-called ground segment of the project is contracted out to the telecommunication company, Telespazio. Via a 10-meter-wide dish antenna in Malindi, Kenya, scientific data from each revolution of the satellite are sent to earth and new instructions from earth are sent to the satellite. A 'regular' satellite connection takes care of the contact between Malindi and Telespazio's ground station in Fucino, about 100 kilometers east of Rome in the foothills of the Apennines.

Not that BeppoSAX was problem-free after the launch. One of the Italian X-ray telescopes broke down and four of the six gyroscopes that are necessary for maintaining position failed during the first two years. But thanks to the clever programmers, the satellite could still be accurately directed by making use of the measurements of a sun sensor,

On April 30, 1996, BeppoSAX was launched with the help of an Atlas G Centaur rocket.

three star trackers and a magnetometer. Money was a bigger problem. In the beginning of 2000 it already looked as though the project would have to be stopped because of budget considerations, and in May 2002 ASI plans to switch off the satellite for good.

But the most important 'trophy' has already been won. On the first floor of the Telespazio building where the BeppoSAX control center is housed, there hangs a framed document of the Bruno Rossi Prize from the American Astronomical Society. In 1998, the BeppoSAX team shared the prize with Jan van Paradijs from the University of Amsterdam for the discovery a year earlier of the afterglows of gamma ray bursts at X-ray wavelengths and in visible light. It was this discovery that brought an end to the uncertainty about the distance of the mysterious gamma ray explosions. BeppoSAX had etched its name in history for all time.

Things weren't going as well with the American High Energy Transient Explorer. NASA had decided to launch the small satellite using a light, commercial booster rocket; their choice was the Pegasus from Orbital Sciences Corporation in Dulles, Virginia. The Pegasus rocket can best be compared with a small cruise rocket. It is carried, satellite and all, under the fuselage of a large cargo plane, and released over the ocean at an altitude of about 13 kilometers. After a free fall of several seconds the three-stage booster rocket is fired and 10 minutes later the satellite reaches an orbit around the earth.

The Pegasus made its maiden voyage on April 5, 1990, and for the first five flights it was completely successful. For the launching of HETE a more powerful rocket called the Pegasus XL was necessary. But in June 1994 that rocket exploded on its first voyage with a military satellite on board. A year later there was another accident and the HETE team really started to worry.

However, the U.S.Air Force was in need of the launch capacity of the new rocket and together with Orbital Sciences they worked very hard to find a solution to the problem. This was at the end of 1995. In March of 1996 the Pegasus XL again would be used to launch an Air Force satellite and George Ricker and his team were told that if the rocket worked well this time, they could send up HETE within three months. But, should anything go wrong, there would be a delay of a good year and a half – an unworkable situation, according to Don Lamb.

The launch on March 8th went according to plan and later in that same year Pegasus XL chalked up two more successful flights; but for Lamb and his colleagues it still didn't sit well. They had worked 13 years on the satellite and it seemed sheer madness to launch with a rocket that had failed on two attempts out of five. NASA turned a deaf ear to their arguments. It was 'put up or shut up' for the HETE team; it either went up with the next Pegasus flight or the project got scrapped.

Incidentally, HETE was not the only 'passenger' aboard the Pegasus XL. There was also an Argentinian satellite, the SAC-B (Satélite de Aplicaciones Científicas). Both satellites were mounted on the third stage of the Pegasus rocket, the SAC-B on the outside and the HETE on the inside of a construction that was especially designed to launch two satellites at the same time. After the ignition of the third rocket stage the two passengers would be disengaged.

On Monday, November 4th, 1996, about a half year after the launching of BeppoSAX it finally happened: after several delays the Pegasus XL rocket called 'Maggie' was let loose over the Atlantic Ocean and set its course for space. According to Ricker, all went well up to and including the ignition of the third stage. Then came an unsettling report from the control room: 'No further third stage pyro events.' Slowly but surely the import of this calamitous announcement forced its way through to everyone. The uncoupling didn't work. SAC-B and HETE sat bound on the burned out rocket stage, the Argentinian satellite on the outside and the gamma ray burst satellite on the inside.

In the beginning there was confusion about what was going on, but observations from the Haystack Radar Observatory left no doubt that there was just one object orbiting the earth, and not three, as had been expected. SAC-B could open its solar panels but because of the ballast from the Pegasus stage and the HETE satellite locked up inside, it wasn't possible to aim the panels in the direction of the sun. And to make matters worse, they were partly in the shadow of the Pegasus. After several hours the energy supply of the satellite was in danger and one day later all radio contact was lost.

And HETE? Lamb has difficulty holding back his tears when he recalls that traumatic event. 'The satellite was programmed to unfold its solar panels at a given time,' he relates. 'And from the telemetric information we received from SAC-B, it appeared that it did so. It was as though we could feel HETE moving in its coffin for four minutes, and then an hour of silence, followed again by four minutes of klonking and banging . . . it was horrible.'

Within a couple of hours the whole thing was over. Without continuous charging from the solar panels the batteries on board the HETE ran dry very quickly. 'It's a tough world,' says Ed Fenimore, 'and you have to learn to live with these set-backs.' But even for Fenimore, with all his experience in space research, this was a blow and for several months he was deeply depressed. For Lamb, who had never before worked on a satellite program, the jolt was overwhelming. 'It was a black day in my life. It took me two years to get over it,' he says, 'it was as though you had lost a child.'

There wasn't much opportunity to work through the loss because Ricker got busy, on the 5th of November, working out a strategy for a re-try. They had backups for most of the instruments and parts of the satellite; it should be possible with very little investment to build a HETE 2. 'For several months we worked continuously,' says Lamb, 'and in February 1997 it was as good as certain that HETE 2 would get NASA's approval.'

But during that same month astronomers from the University of Amsterdam, using observations from BeppoSAX, saw for the first time the afterglow of a gamma ray burst. And it wasn't long before it was irrefutably shown that gamma ray bursts are billions of light years away, in the depths of the universe. Lamb, who, two years before in his debate with Paczyński, had invested all his hope in HETE, was not in the mainstream of the action now; and what was even more discouraging was that he also had to give up his treasured neutron star model. For years there had been room for endless debate about gamma ray bursts, but now it was the real thing.

5 Beaming in on afterglows

'There is no such thing!' Alice was beginning very angrily, but the Hatter and the March Hare went 'Sh! Sh!' and the Dormouse sulkily remarked, 'If you can't be civil, you'd better just finish the story for yourself.'

John Heise loves puzzles. For about the past ten years he has immersed himself in the study of Akkadian, a Semitic language from old Mesopotamia that was, for a few thousand years, the *lingua franca* of that land between the Euphrates and the Tigris. On Sunday mornings, while his girlfriend is absorbed in working out the crossword puzzle in the newspaper, Heise, surrounded by fat dictionaries, is steeped in a cuneiform text trying to find the grammatical base of another language group and the distinction between logograms and phonograms. He designed a cuneiform font for his own use on his PC and in his free time he even put together a complete website on the Akkadian language.[1]

Heise is neither a professional linguist nor an archeologist; he is the project scientist of the WFCs – the Wide Field Cameras for BeppoSAX. It was with these cameras that the big breakthrough in the field of gamma ray burst detection was made. During a press conference shortly before the launch of BeppoSAX, held at the Utrecht laboratory of the Space Research Organization Netherlands (SRON), Heise told the journalists about the great potential of the cameras, which were originally designed to study variable X-ray sources in the Milky Way galaxy. While the search for gamma ray bursts in the eyes of most of the Italian team members was a side issue, Heise ventured to say that the Wide Field Cameras might possibly be able to pick up gamma ray bursts, perhaps eight times a year, and then, within a couple of hours, determine the location. A fast fix on its position would also make it possible to observe the source of the explosion with large optical tele-

[1] http://saturn.sron.nl/~jheise/akkadian/index.html

(*Left*) *John Heise*, the project scientist for the Wide Field Cameras on board the BeppoSAX satellite.

(*Right*) *Jan van Pardijs* led the team that discovered the first optical afterglow in 1997. That discovery made it possible to determine the distances of the mysterious explosions.

scopes. This could eventually lead to the solution of the distance dilemma.

All the gamma ray detectors, from the simplest instruments on the Vela to the colossal detectors on BATSE have one great shortcoming: they cannot accurately determine the exact position in the sky where the gamma ray bursts are occurring. Working with different detectors, all facing in different directions, it is possible to tell from which general direction the burst is coming, but it had not yet been possible to pin it down precisely. It is as if you are sitting blindfolded in the sun and you have to determine its direction by the difference in the warmth you feel between your left and right cheek. Even BATSE with it eight sensitive detectors could only locate the position of the gamma bursts to within a few degrees of accuracy, at best.

But gamma ray bursts also produce less energetic X-rays, as had been observed by the Japanese X-ray satellite Ginga. This is not surprising: just as large earthquakes are accompanied by lighter shocks, and violent rainstorms also send down smaller drops, you can expect that during a gamma ray burst, X-rays also come free. If you can observe these X-rays at the same time, the fixing of the position of the gamma ray bursts in the sky is much more accurate, since an X-ray telescope enables you to take a real picture of the sky.

Unfortunately, gamma ray bursts come unexpectedly so you don't know in which direction to face your camera. The only thing you can do is cover as great a portion of the sky as possible with X-ray cameras and wait until a gamma ray burst comes into view of one of them. The greater the field of view of your camera, the more often a burst will be caught.

For Heise the problem wasn't so difficult. Each of the Wide Field Cameras has a field of view of 40×40 degrees (although their sensitivity decreases toward the edge of the field); this can be compared with what one sees through a portrait lens of a photo camera. Together, they cover an area of many hundreds of square degrees – a few percent of the whole starry sky. Considering that around 400 gamma ray bursts are observed each year, generally speaking it can be expected that a handful of those gamma ray bursts will be picked up by the X-ray cameras.

When a conventional photographer speaks of a picture angle of 40 degrees, he is not talking about a wide-angle lens, but for an X-ray camera it is indeed sizable. Most X-ray cameras and telescopes have a much smaller field of view. This has to do with the unusual demands made upon the optics. To make a picture of your surroundings in visible light, you use lenses or curved mirrors to bring the rays of light together to one focal point. But high-energy X-rays just transverse any mirror or lens – there is no talk of reflection or refraction, and therefore, there is also no picture.

The only way to focus X-rays is to make use of 'grazing incidence'. When an X-ray photon falls at a very small angle on a mirror-smooth metal surface it is reflected, in the same way that a flat pebble skims over the water. For many years all X-ray telescopes were built to take advantage of this phenomenon. They were made of various grazing incidence mirrors, nested inside each other. It is a wonderful way to make X-ray photos of the sky but it has one drawback: because the angle

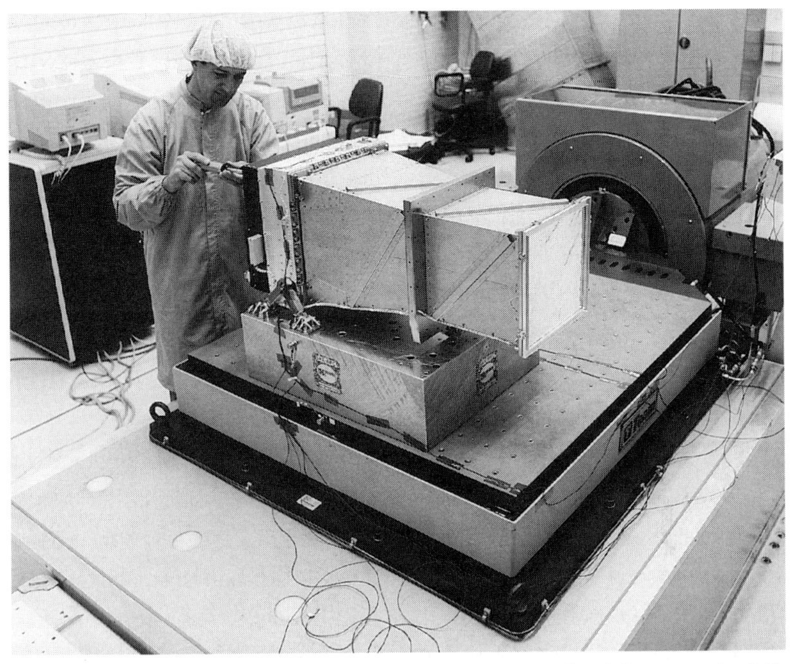

One of the Wide Field Cameras from BeppoSAX. Behind the protective foil on the right is the coded mask.

of incidence must be very small, such a telescope, of necessity, has a very small field of view. Wide Field Cameras cannot be made in this way.

Heise's X-ray cameras work on a completely different principle. No optics are involved. In fact, they are really advanced pinhole cameras. An ordinary pinhole camera (*camera obscura*) consists of a light-proof box with a small opening at the front. Through this opening the outside world is projected upside down onto the back wall of the camera. A ray of light that comes from the upper left, via the hole, ends at the bottom right of the camera. In the middle of the back wall there are only the light rays that come precisely from the viewing direction of the camera. In this way, without lenses and mirrors, you can make a very simple camera. Naturally, the big disadvantage is that so very little light can come through the hole that a long exposure time is necessary. You can make the hole bigger and expose for a shorter time but then the picture becomes less sharp.

A pinhole camera has the same disadvantage for X-rays: if a bright X-ray point-source comes into view, you can get an accurate fix on it with a pinhole camera. But because so few X-ray photons come in through the little hole, it takes a long time before you can detect the source with any certainty. Incidentally, detection does not occur with a photographic film, as in a regular camera, but with a gas tank filled with noble gas in which hundreds of extremely thin tungsten wires are strung in a mesh-like pattern. An incoming X-ray photon produces an electrical charge, and by measuring the charge on the horizontal and vertical wires, the position of the photon can be measured with a precision to within half of a millimeter.

To detect more X-ray photons you could make the hole bigger but again, you pay the price of losing the sharpness of the image. However, there is still another way: you can make more holes. Each hole creates its own image of the X-ray source and if all the different signals are added up, the sensitivity of the camera increases. A camera with ten holes will detect a particular X-ray source ten times as fast as a camera with only one hole.

The Wide Field Cameras of BeppoSAX work according to this principle. The square metal plate on the front of the camera measures 25 centimeters across and has no less than 21,845 holes, each measuring one square millimeter. One out of three X-ray photons that fall upon this 'coded mask' is let through. Each X-ray source in the sky is imaged many thousands of times on the surface of the detector. If you bear in mind that in an area of 40 by 40 degrees there are often many X-ray sources, then it is no surprise that the observational data can only be transformed into 'real' images with the help of extensive computer interpretation.

The idea of a coded mask camera for X-rays had already been suggested back in the 1960s and the first successful flight was carried out with a British sounding rocket in 1976. Ed Fenimore at the Los Alamos National Laboratory in New Mexico helped to improve the concept. Fenimore realized that the trick would not work when thousands of holes are systematically distributed over the coded mask because the computer is then unable to derive the positions from the exposure pattern unambiguously. Nor does an arbitrarily random distribution work. Instead, there had to be a 'pseudo-random pattern'. In the beginning, Fenimore thought the idea was unworkable but when he began to think it out scientifically, he hit upon a solution. It wasn't too long before he patented his creation. Not that he made much money out of it. There were very few coded mask cameras built in the world and Fenimore would have to be given a share of the profit only if anybody wanted to build and sell them.

The coded mask cameras built by SRON in Utrecht are, in any case, *not* for sale. The space research lab has built up a worldwide reputation in the area of X-ray detectors and spectroscopic gratings for the sensitive X-ray spectrometers of NASA's Chandra X-Ray Observatory and the European XMM–Newton Observatory, the two largest X-ray satellites ever built.

The Wide Field Cameras for BeppoSAX were not the first coded mask cameras built by SRON. In the early 1980s the Dutch had seriously planned to build a third national satellite, after the ANS (Astronomical Netherlands Satellite) and the IRAS (Infra-Red Astronomical Satellite), which had been built in cooperation with the United States and Great Britain. The third satellite would be an X-ray satellite, equipped with a great number of coded mask cameras that could keep an eye on a very large part of the sky. In this way it would be possible to track and register short-duration X-ray explosions and rapid flashes from X-ray sources.

Heise, who, at that time, had just graduated from the University of Utrecht, can well remember the plans that were being made for this TIXTE (TIming X-Ray Transient Explorer). It wasn't too long before it was recognized that the project would be too expensive. TIXTE never materialized but in the meantime one coded mask camera was already being built in Utrecht. It would be a shame if this lay unused on the shelf. Fortunately, Kees de Jager, founder and former director of SRON, had good connections with astronomers in the Soviet Union. Through Rashid Sunyaev from the Russian Academy of Sciences in Moscow, de

Jager managed to get the camera on board the Russian space station Mir.

And thus COMIS came into existence – the COded Mask Imaging Spectrometer. It was launched in 1997 on board the scientific Kvant module of the Mir. According to Heise, COMIS, which orbited the earth until Mir's descent in March 2001, can be seen as the forerunner of the Wide Field Cameras used by BeppoSAX, even though its effective field of view is just 8×8 degrees. Although COMIS did observe a number of X-ray sources, mainly in the direction of the center of the Milky Way galaxy, its scientific output was relatively limited. Heise's colleague Jean in 't Zand, whose thesis was based on the COMIS project, said that over the years scarcely more than a total of two weeks of worthwhile observations were carried out.

And there was one major reason why it was such a challenge to make good observations. COMIS was immovably placed on the outside of the Kvant module. In order to get a particular X-ray source in view the whole space station had to be turned. Moreover, the astronauts had to be as still as mice because the smallest motion could cause unwanted vibrations. But to look on the rosy side, it turned out to be good that COMIS was on a manned spaceship because shortly after the launch, a part of the detector got out of order, and during two spacewalks in 1988 the Russian astronauts completely replaced the detector on the X-ray camera with a spare one.

Even though seven new X-ray sources and fourteen X-ray explosions were discovered, the real importance of the instrument, according to in 't Zand, lay in the experience that was gained in the calibration of the camera and the software for the data analysis. In this last respect, it is a puzzle that does not take a back seat to the deciphering of a Babylonian clay tablet. The software of the camera must be able to determine the position and energy of the incoming X-ray photon on the basis of extremely small voltage differences on the tungsten wires in the detector. And the software on earth must plow through the mounds of positional data to reconstruct an X-ray 'photo' of a piece of the sky, based on a pseudo-random pattern of holes from the coded mask – a process that very appropriately is called deconvolution.

It is no wonder that Heise has not had the time to work on his Akkadian website since the spring of 1996. Although his colleague Rieks Jager had supervised the construction and launch of the Wide Field Cameras (WFC) – a job that took him some ten years – Heise, who would succeed Jager as project scientist in the spring of 1997, was heavily involved in the project as well as having puzzles on his mind. Shortly after the launch, all the instruments on board BeppoSAX had to be tested and calibrated, and while that task was in full swing, the results of the first preliminary measurements were beginning to come in. The Italian group in Rome was, of course, concentrating on the Italian instruments, but the astronomers in Utrecht knew all too well that they needed to analyze the WFC observations and study them as quickly as possible. So SRON sent two WFC scientists to the control center at Telespazio. Hans Muller and Michael Smith were to be the Roman eyes, ears and hands for Heise's team in Utrecht.

Muller's impression of the first few weeks in Rome was one of sheer chaos. The Italians were completely focused upon the LECS and the MECS instruments[2] which were the X-ray telescopes for low and medium X-ray energies whose small field of view formed the 'tele-lenses' of BeppoSAX. Nobody really had time to give any attention or thought to the Wide Field Cameras from Utrecht, or the role they might play in unraveling the mystery of gamma ray bursts. It was barely mentioned. To make matters worse, there were problems with the data processing. The Quick Look Analysis Team, which had to evaluate the incoming data of each satellite orbit, did not have the software to turn the data into X-ray images, and the deconvolution, which was partly done by Muller's colleagues in Utrecht, took much more time than expected.

At the same time there was considerable friction between the Dutch

[2] LECS and MECS stand for Low Energy and Medium Energy Concentrator Spectrometer.

One of the control rooms for BeppoSAX in the Telespazio building in Rome. Standing on the right is the project scientist, Luigi Piro.

and Italian astronomers, which didn't make things easier. What led to the controversy was the question of right of ownership of the observations from BeppoSAX. The X-ray satellite was set up from the very beginning as a real space observatory, and like the telescopes on earth, astronomers from all over the world could submit their observation proposals. When they are approved of by a special time allocation committee, the observational data that are collected become the exclusive property of the observer for a given number of months. This gives the observer ample opportunity to analyze them and to write up his or her research.

All X-ray detectors and telescopes on BeppoSAX, including the Utrecht Wide Field Cameras, were made available to the international community in this way. For some time before the launch, all astronomers were encouraged to submit observation proposals. But the science steering committee for BeppoSAX had actually overlooked one important instrument – Enrico Costa's Gamma Ray Burst Monitor (GRBM). With no announcement of opportunity, no call for proposals,

Costa's instrument remained a 'private instrument', and the observations from the gamma ray burst monitor remained, in principle, the property of Costa and his fellow team members for all time.

John Heise realized early on that he could not do without Costa's team if he wanted to chase down gamma ray bursts with the Wide Field Cameras. Such a search would only make sense if you knew precisely the time at which the gamma ray bursts were observed. After that you could look at the measurement data from the WFC's and then see if you were able to find X-rays to match. If that worked, then the X-ray source could be studied in detail with the BeppoSAX 'telelenses'.

As soon as observation proposals could be submitted for the X-ray telescopes, Heise was in on it. Together with the principle investigator, Jan van Paradijs, an experienced high-energy astrophysicist from the University of Amsterdam, he sent in a proposal to use this method to catch the X-rays from the gamma ray bursts. And to make certain that the essential data from the gamma ray burst monitor were available, Costa's colleague Graziella Pizzichini from the TeSRE institute in Bologna (Istituto Tecnologie e Studio delle Radiazioni Extraterrestri) was invited to join the Dutch team.

Van Paradijs and Heise were a little premature with their proposal, according to the allocation committee from BeppoSAX. The fact was that this round of proposals was for the X-ray telescopes and the Dutch team could not yet specify which objects they wanted to observe. First, a gamma ray burst would have to be found in the observational data of the Wide Field Cameras, and since another separate round would come for WFC proposals, the proposal from van Paradijs and Heise was shoved aside.

But Pizzichini told her colleagues in Italy about the rejected proposal, much to the chagrin of van Paradijs. As a result, when it came time for the Wide Field Cameras' observing proposals to be handed in, there lay on the table not only the Dutch proposal but also an almost identical one from the Italians with Costa as principle investigator. Costa's proposal was accepted, the one from van Paradijs was not. Should the gamma ray detector on BeppoSAX register a gamma ray burst, then the Italians were to be the first to search through the observations from the Wide Field Cameras, and if they were able to determine accurately the position of the burst in the sky, it would immediately become the 'property' of Costa and his colleagues. It seriously began to look as if the first identification of a gamma ray burst would be an Italian discovery.

Van Paradijs could not sit back and accept this situation. Costa's GRBM was not the only instrument up in space that could register gamma ray bursts. In cooperation with Gerald Fishman and Chip Meegan from the Marshall Space Flight Center a proposal was submitted for the next WFC round which would make use of the BATSE detections. As soon as BATSE picked up a gamma ray burst in a part of the sky that, by chance, lay in the viewing range of the Wide Field Cameras at that moment, van Paradijs would look through the WFC data for matching X-rays. This proposal was accepted, although it did not go into effect until 1997.

In the summer of 1996 Costa's team really had a monopoly on information. It so happened that Costa's team member, Marco Feroci, came to Hans Muller in August with a list of about 25 dates and times. Feroci had been looking through the observational data of the GRBM for detections that might possibly have been caused by gamma ray bursts. This was a mammoth task because 99 out of 100 detections are the result of either electrically charged particles in the radiation belts of the earth, explosions on the sun, or voltage fluctuations in the detector itself. Undoubtedly, there were still many of those 'false' detections on Feroci's list.

In his role as an external member of the Utrecht-based WFC team, Muller had to find out if the Wide Field Cameras had, at the corresponding moment, observed anything unusual. By then, all the measurement data were on tape, and in principle, it was simple to see if, at that particular moment, there were also high levels of X-rays detected. If one of the two cameras had registered an X-ray peak precisely at the moment that the GRBM caught a blast from a gamma ray burst, then one could say with certainty that a gamma ray burst was picked up in the field of view of the X-ray camera, and by deconvoluting the WFC measurements to bring forth a real image of the sky, it would be possible to determine the position of the gamma ray burst fairly accurately.

Nevertheless, it was still a time-consuming job, and after more than 15 fruitless attempts, it became frustrating. But Feroci and Muller didn't give up. Maybe some of the detections on Feroci's list were 'false', and others probably fell outside the field of view of the Wide Field Cameras, but still there was the possibility that at least one real gamma ray burst had taken place in the right part of the sky! Take for instance, the detection on Saturday, the 20th of July, 1996, at 11:37 Universal Time. This one had all the earmarks of a bona fide gamma ray burst, and moreover, since it was observed by only one of the four GRBM detectors, it must have appeared from the side. Since the Wide Field Cameras of BeppoSAX look exactly sidewards, the chances were quite good that this blast had indeed been seen.

Sure enough, in the WFC data, Muller found that an X-ray burst had occurred around the same time as the gamma ray burst. Could this be the long-awaited X-ray counterpart? To find out, it was necessary to know precisely the time of the X-ray burst, and this was a difficult task. It involved the exact calibration of the clock on board BeppoSAX. Muller was exhausted after working continuously for many days, and when he didn't manage to find a convincing correspondence between the timing of the X-ray burst and the gamma ray burst, he went to bed.

Meanwhile, Feroci learned that the BATSE detectors on the Compton Gamma Ray Observatory had also observed a strong gamma ray burst on July 20th. BATSE had determined the sky position to within a few degrees and it was practically in the center of the field of view of the Wide Field Cameras. There could be little doubt that the X-ray peak detected by the WFCs was the counterpart of the gamma ray burst. Working from the same WFC data, Feroci and his colleague Mauro Orlandini finally succeeded in calibrating the clock and at once, it was apparent that they had a real hit: on July 20th, exactly at 11:37 Universal Time, one of the two Wide Field Cameras of BeppoSAX had caught a short, powerful blast of X-rays. The next day, using the precise timing of the X-ray burst, Muller was able to deconvolve the WFC data into an X-ray image of the sky, and to determine a preliminary sky position.

Unfortunately, it was already August the 13th, a good three weeks after the cosmic explosion. Is it possible that a weak X-ray source could still be seen at the position of the gamma ray burst? Could a large optical telescope or a radio telescope find something? At SRON in Utrecht the WFC observations were analyzed further so that the sky position of the burst could be determined even more accurately. It turned out that the high energy radiation came from an area in the far north of the constellation Hercules, right under the head of the Dragon.

Never before was the position of a gamma ray burst so quickly and so accurately determined. Via the coded mask technique the direction from which the burst came could be traced with an accuracy of around 10 arcminutes – a third of the apparent diameter of the full moon. The Inter-Planetary Network was able to give results that were even better, but not in so short a time. Who knows? It may finally have come to pass that the source of a gamma ray burst could be identified.

An arbitrary little area of the sky with a diameter of 10 arcminutes contains an unimaginable number of faint stars and distant galaxies, and a quick look at the *Digitized Sky Survey*, a photographic atlas of the sky available on the Internet,[3] did not show anything unusual. A search with an optical telescope was a hopeless job; there were too many candidates. After discussing it with Luigi Piro, the BeppoSAX project scientist, the satellite was turned so that the area in question could be studied with the X-ray telescopes. But even with those sensitive instruments, there was nothing to be seen.[4] Nor did a search requested by Costa with the colossal Very Large Array radio telescope in New Mexico, turned in the right direction, yield more information.

[3] http://archive.stsci.edu/dss/dss-form.html

[4] On September 30th an IAU circular reported that an X-ray source at the position of a gamma ray burst was observed on September 3rd with the MECS telescope. It later turned out that this source was not related to the burst on July 20th.

Whatever it was that exploded in Hercules was completely burned out in a couple of weeks.

On September the 3rd, well before the unfortunate launch of the American High Energy Transient Explorer, Piro, Costa, Feroci and colleagues sent out an electronic circular via the International Astronomical Union (IAU) announcing to the world the news of their discovery of X-rays from GRB 960720 (GRB stands for gamma-ray burst; the number code gives the date). From one day to the next, BeppoSAX changed from an obscure Italian X-ray satellite into a promising gamma ray burst observatory. Although the burst that took place on July 20th was extinguished before it could be studied in depth, it was obvious that if one were able to act more quickly it could lead to a successful identification. If the follow-up observations could be done within a day instead of a month then the positive identification of the gamma ray burst would have to happen. Slowly but surely the net was tightening around the mysterious cosmic explosions.

In the second half of 1996 the BeppoSAX team did everything possible to react quickly to a gamma ray burst. As soon as the GRBM picked up a burst, the X-ray observations from the Wide Field Cameras were minutely studied and when one of the two cameras registered a burst at the right moment, the position could be known within a day so that BeppoSAX could turn its sensitive X-ray telescopes in the right direction.

Meanwhile, Costa's team had contacted observatories in Italy and Spain and offered to pass on information about burst positions as soon as possible, so that within a 24-hour period they could also be searched for with optical telescopes. If a few photographs, one after the other, were made, then perhaps the point at which the explosion took place would reveal a fading light source. There was also close cooperation with Dale Frail from the National Radio Astronomy Observatory (NRAO) in Socorro, New Mexico, and with Shrinivas Kulkarni from the California Institute of Technology in Pasadena. Frail had access to the Very Large Array (VLA), one of the most powerful radio telescopes in the world, and if he succeeded in finding a radio source at the position of a gamma ray burst, its position in the sky would be known with a high degree of accuracy. Kulkarni could then make optical observations with the 10-meter Keck telescope in Hawaii, the biggest telescope in the world.

There were enough gamma ray bursts in the last months of 1996, but the Wide Field Cameras kept looking in the wrong direction. Together, they did not cover more than 2% of the sky which means that on average, only one out of fifty bursts would come into view of the cameras. It was not until the 11th of January, 1997, that BeppoSAX finally got a bite. On that Saturday at 09:44 Universal Time, an extremely powerful flash was registered which lasted almost a minute and showed several distinct peaks – a hallmark of gamma ray bursts.

The observations were not immediately available because BeppoSAX only has contact with the ground station in Malindi once every 100 minutes. But as soon as the Science Data Center got the information, the measurements from the Wide Field Cameras were quickly analyzed. A hit! Precisely at the right moment WFC 2 had registered a large amount of X-rays. At once, the scientists and technicians on duty contacted Costa and Piro, and on the same day Heise received a call asking him to figure out the sky position, using the information from the WFC. This was a skill the Italians had not yet mastered.

Less than 24 hours after the explosion, Costa could call through a rough estimate of the position to Frail: the gamma ray burst came from the constellation Serpens; the error box had a diameter of around 20 arcminutes. A short time later the 27 dish antennae of the Very Large Array began to scan the skies. But there were others busy doing the same thing. Heise sent the position to about 100 interested astronomers, including van Paradijs who, with his graduate student, Titus Galama, went to search for the radio source with the radio telescope at Westerbork, the Netherlands.

On the 12th and 13th of January the suspect piece of sky was studied in great detail with the X-ray telescope of BeppoSAX. In that area they found two faint X-ray sources. One of the two might possibly be related to the gamma ray burst. If they could also get a confirmation from the Very Large Array, then the identification would be complete.

Frail had decided to let the Italians know that he was the man they needed to make this kind of observation. The Very Large Array is twice as sensitive as the Westerbork telescope and, as staff astronomer, Frail knew the instrument through and through. If there was anybody in the world who could observe the afterglows of the gamma ray bursts at radio wavelengths, he was the one.

The VLA observations revealed several radio sources, but only one of them was in the 20 arcminute error box determined on the basis of the measurements of the WFC. This one radio source coincided precisely with one of the two faint X-ray sources from BeppoSAX. Could this be the object that produced the gamma ray burst on January 11th? Or did the X-ray and radio sources have nothing to do with the gamma ray burst but, by pure coincidence, happened to be in the same direction in the sky?

Frail did not take anything for granted. If this was a chance coincidence, then nothing unusual would be found with the radio source. But if this was the afterglow of the gamma ray burst, you would expect that, with the passage of time, it would become more and more faint. For a few weeks Frail kept a careful eye on the faint radio source and sure enough, the radio intensity diminished! Kulkarni, in the meantime had observed the object with various optical telescopes, including the 10-meter Keck telescope. From the spectrum it appeared without a doubt that this was a so-called BL Lacertae object – a very distant galaxy with an extremely bright nucleus.

Frail and Kulkarni were completely convinced that they had solved the mystery of the gamma ray bursts. BL Lacertae objects are known for their remarkable activity. They send out many X-rays and radio waves and their luminosity keeps changing. The energy that comes from such an active galaxy is probably produced by a supermassive black hole in the galactic core. Why shouldn't there also be gamma ray bursts in the immediate vicinity of such a black hole? Moreover, BL Lacertae objects are extremely rare and the chance that, by coincidence, one would be found in that small area of the sky with a diameter of 20 arcminutes is very small.

There seemed to be no flaw in their reasoning. Gamma ray bursts apparently come from active galaxies at a distance of hundreds of millions of light years away from the earth. Frail and Kulkarni got the first definite identification. It was high time to publish in the scientific weekly journal, *Nature.* The article was written quickly, offered to *Nature*, submitted for peer review, and after receiving positive reactions it was accepted for publication. Sometime in March the news would be made known.

But Frail had not figured on having to deal with Jean in 't Zand, who, after working as a postdoc with Ed Fenimore at Los Alamos and as a gamma ray spectroscopist at the Goddard Space Flight Center, had come back to the SRON laboratory in Utrecht in January 1997. There, he gave his full concentration to calibrating the Wide Field Cameras of BeppoSAX, which, in his opinion, could be much better. The computer software that deconvolved the WFC measurements provided sky positions with an average accuracy of 10 arcminutes. But with the experience he had acquired with the COMIS camera on board the space station Mir, in 't Zand felt more strongly than ever that there was room for improvement.

By the end of February he had partially rewritten the software for the analysis and it was now possible to determine the positions in the sky with an accuracy of 3 arcminutes. Through an IAU Circular the improved positions were made known to the world, to the great consternation of Frail and Kulkarni. Those damn Dutch had reduced the error box and suddenly it seemed that the suspected BL Lacertae object lay *outside* of it!

Frail was at his wits end. Was in 't Zand really sure? Of course he was. Now it was impossible that the galaxy Frail had been so sure of had anything to do with the gamma ray burst. But the article heralding the discovery had already been accepted by *Nature*! What a loss of face if, after it was published, it turned out to be wrong! Frail and Kulkarni contacted *Nature,* explained the situation and withdrew the article. Although the *Nature* editors were completely understanding and certainly did not spread rumors about the affair, nevertheless, word got out and it was a humiliating situation, according to Frail.

For the second time, the Wide Field Cameras of BeppoSAX had observed both X-rays and gamma ray bursts, and for the second time they did not succeed in identifying the source of the burst, even when the position in the sky was known within the same day. The frustration the Italians were feeling at being so close, and the panic Frail and Kulkarni experienced with their *Nature* article, showed the extent to which many astronomers had invested their knowledge and their emotions in solving once and for all the most obstinate enigma in modern astrophysics. Moreover, the message from GRB 970111 was clear enough: you don't trifle with gamma ray bursts. Anybody who wants to catch them had better be there on the spot.

In the meantime, the situation between Costa and Heise worsened after January 1997. According to Costa it was completely unethical for Heise to have given van Paradijs the position he had from GRB 970111 before an IAU circular had been released. Not that anything was discovered with the Westerbork telescope, but the fact remained that van Paradijs had all the information about the position made available to him before Costa had given his permission, and this was done while the position of the burst was Costa's 'property'. Heise countered with the argument that everyone profited when the positions were made public as quickly as possible for the simple reason that the chance of solving the mystery was greater. And what is more, it was *his* X-ray camera and the positions were determined by *his* team. If there was anyone who was acting unethically, according to Heise, it was Costa.

After a few months the altercation was laid to rest. 'I found out that John is an expert in the Akkadian language,' says Costa, who himself has a great interest in archeology. 'Anybody who can read cuneiform, can't be all bad, I said to myself. We talked the whole thing out and I am convinced that he meant no wrong.'

But that was many months and two identified gamma ray bursts later. The Dutch–Italian relationship had hit bottom when Jan van Paradijs's Amsterdam team had (illegally, according to some Italians) become the first to have made an optical identification of a gamma ray burst which triggered the long awaited revolution in that field of research. The cosmic super explosion on the 28th of February 1997 not only created spectacular fireworks in the universe, it also set off a brawl that reverberated throughout the earthly world of ambitious researchers.

6 First among equals

'All right', said the cat; and this time it vanished slowly, beginning at the end of the tail, and ending with the grin, which remained sometime after the rest of it was gone.

That evening, in the lobby of the Hilton Hotel in Huntsville, Alabama, almost all the female employees were standing around waiting in excited anticipation. You don't come across a group like that often. The attraction certainly wasn't the couple of hundred astronomers who were walking around there for the 5th Huntsville Gamma Ray Burst Symposium, made up mostly of indifferently dressed men who seemed to be in their own world and kept themselves busy from morning to night with unintelligible, far out problems. No, the excitement had to do with the basketball teams that had come to town for a big game and were spending the night at the Hilton. Handsome, muscled young guys, about 7 ft tall with big sport bags slung over their shoulders and shoes as big as your forearm. A forest of Goliaths passed through the reception hall, real stars with their heads literally in the clouds.

Jan van Paradijs would have felt at home in the middle of these sports heroes some years ago. The Amsterdam astronomer had always been a talented, fanatic basketball player, and what he lacked in height he made up for in skill and staying power. But van Paradijs would have missed this parade of giants. As chairman of the scientific program of the gamma ray burst symposium he would never skip a lecture, but at night he would always sleep at home in a suburb of Huntsville rather than in the hotel. For the past six years he had divided his time between the University of Amsterdam and the University of Alabama, and Huntsville had become his second home.

But this time, Jan van Paradijs was not present. In October 1999, worn out and weary, he lay in the hospital in Amsterdam being treated for cancer. By telephone he spoke to his colleagues and wished them a successful conference. A week later he received a book full of personal wishes and he listened as they told him the latest scientific news.

It was the fifth time that Gerald Fishman from the Marshall Space Flight Center had organized a symposium on gamma ray bursts in Huntsville. It had become a bi-annual tradition that attracted a variety of scientists from all over the world: high-energy astrophysicists, supernova experts, theoreticians, observers, satellite builders and cosmologists. But according to Fishman this fifth symposium may well be the last. The research on gamma ray bursts had grown up and in the future there would probably be more of a need for meetings devoted to specific aspects of research.

By 1999 the riddle of the distance scale of gamma ray bursts was already solved. And as the tempo of research picked up, there were credible indications that the cosmic explosions could be the birth cries of black holes. A new generation of satellites was rapidly evolving, intent on pulling back even further the veil of mystery surrounding those elusive explosions. Discovery and exploration had bowed to research and quantification; groping the heavens for clues had given way to focused study.

The name of van Paradijs will forever be bound up with the revolutionary breakthrough that helped to make this development possible. In February 1997, his team was the first to succeed in detecting a gamma ray burst in visible light, and finally it became possible to pinpoint accurately the location of one of those searing explosions – thirty years after the first observations by the Vela satellites. 'This could be a turning point in GRB astronomy,' wrote van Paradijs and his colleagues in their *Nature* article of April 17th, 1997. A more profound understatement would be hard to think of.

Gamma ray bursts sometimes last no longer than several tenths of a second and the very longest never remain for more than a couple of minutes. Perhaps there is visible light sent out during an actual burst. But it is practically impossible to direct an earth-bound telescope to the right position in the sky in so short a time, particularly when you remember that a detector like BATSE cannot determine the position accurately. Van Paradijs had not set his sights on observing the burst itself but rather, on finding its afterglow.

In 1993, Bohdan Paczyński and his student James Rhoads had already predicted that such a high-energy explosion as a gamma ray burst would have afterglows at radio wavelengths and they would last for some time, regardless of the exact cause of the explosion. If something with that much energy exploded, reasoned Paczyński and Rhoads, there would have to be material blown into space at extremely high speeds. The shell of such an explosion would collide with the diffuse interstellar material and the resulting shockwave must produce radio waves. If gamma ray bursts occur in far distant galaxies, as Paczyński was convinced they did, the afterglow might be visible for weeks or even months.

Shortly after the gamma ray burst of January 11th 1997, van Paradijs and his graduate student Titus Galama decided to hunt down its radio afterglow. They made use of the most powerful radio interferometer in Europe, the Westerbork Synthesis Radio Telescope in the Netherlands.[1] The WSRT consists of fourteen dish antennae, each with a diameter of 25 meters, placed in a 3-kilometer long east–west line. The interferometry technique involves receiving signals that are caught by separate dishes which are very precisely coordinated with each other so that, in the end, a radio picture is produced that is as sharp as one from an imaginary giant dish with a diameter of 3 kilometers.

During that January, the Westerbork dishes were trained six times on the suspect area in the constellation of the Serpent, each for a period of 12 hours. Nothing was found. If anything were out there at radio wavelengths, it must be very faint. But might it be possible that an afterglow would be visible at optical wavelengths? After all, nothing was known about the mechanism of gamma ray bursts. Why not look for something observable in visible light as well as looking at radio wavelengths?

Looking for a faint visible object at the position of a gamma ray burst

[1] An earlier attempt to find a radio afterglow on March 1, 1994, was unsuccessful.

The Westerbork Synthesis Radio Telescope in the Netherlands was used in a fruitless search for radio waves from gamma ray bursts.

is nothing new, astronomers have been doing that for many years. But the big difference between then and now is obvious. In the past, after intensive analysis of the measurements from the Inter-Planetary Network, it took weeks or even months to determine the exact positions in the sky. But thanks to the Wide Field Cameras on BeppoSAX, astronomers could now have the information they need within 24

The first images of an optical afterglow were made with the William Herschel Telescope on La Palma.

hours. Van Paradijs hoped that those cameras would also make it possible to locate an optical counterpart before it became too faint to observe.

It didn't happen quickly. The University of Amsterdam does not have a telescope of its own; the Netherlands is so densely populated and there is so much light pollution that a professional optical observatory makes no sense. Dutch astronomers have to go elsewhere – and everything takes time. In order to use the 4.2 meter William Herschel Telescope (WHT) at La Palma in the Canary Islands, an observing proposal would have to be submitted. The WHT is part of the British–Dutch Isaac Newton Group, a conglomeration of three optical telescopes at the Observatorio del Roque de los Muchachos, and was the biggest telescope available to the Dutch on relatively short notice.

Van Paradijs, who had devoted more than 20 years to optical counterparts of X-ray sources, left the compilation and analyzing of optical observations to his student Titus Galama. But Galama was trained as a radio astronomer and had little experience with optical data analysis. Fortunately, another one of van Paradijs' graduate students, Paul Groot, did have the necessary experience, and so he and Galama went off together to hunt down the optical afterglow.

With the help of René Rutten, the Dutch director of the Isaac Newton Group, the proposal from Galama, Groot and van Paradijs was accepted rather quickly. But it still took until the 27th of February before John Telting, a staff astronomer at the William Herschel Telescope, could carry out the observations. According to the plan, Telting would take photographs for two nights in succession of the place where GRB 970111 had been seen. It was hoped that in the middle of the sea of thousands of ordinary stars it would be possible to detect an optical afterglow that would reveal itself by being fainter on the second night than on the first.

As it later turned out, it was to no avail. Just one day after the gamma ray burst actually occurred, on the 12th of January, Spanish astronomers photographed the same suspected area; and when they compared it with control photos taken a month later on the 10th and 11th of February, they found no sign of a fading object. The fact was that Alberto Castro-Tirado and Javier Gorosabel from the LAEFF Institute (Laboratorio de Astrofísica Espacial y Física Fundamental) in Madrid had the sky coordinates from Enrico Costa in Rome but did not make their results public until the second half of March. So even though the astronomers in Madrid already had a non-detection by the end of February, it was not known to the Amsterdam astronomers.

On Friday February 28, 1997, the 'Anton Pannekoek' astronomy institute at the University of Amsterdam was a madhouse. In the morning the first observations from La Palma came in and Groot and Galama started to analyze the data. But around 11:00 o'clock they got an e-mail from Jean in 't Zand from Utrecht. One of the Wide Field Cameras of BeppoSAX had, the night before, at 02:58 Universal Time, observed a gamma ray burst that lasted about 80 seconds, in the northwest of the constellation Orion. So it seemed that here again was a new opportunity to find an afterglow, either at radio wavelengths or with an optical telescope.

On that same night, Jean in 't Zand had been roused out of his bed by a call from Costa. The WFC measurements were quickly sent to Utrecht, and with the help of in 't Zand's software the position of the gamma ray burst in the sky was precisely located to within a few arc-minutes. On the basis of their calculations, BeppoSAX was turned in the right direction so that the area from which the burst came could also be observed with the X-ray telescopes on the satellite.

But there was one problem. After the experience with Costa about giving out the sky position from GRB 970111, Heise didn't dare give a gamma ray burst position to his Amsterdam colleagues again. 'We even hesitated to call van Paradijs', says Heise. 'Finally we decided that we would report that a new gamma ray burst had been observed, but we would refrain from giving its exact position.' If the Amsterdam group wanted to track down an afterglow, they would have to make contact with Piro and Costa themselves.

But van Paradijs wasn't in Amsterdam at all. He was asleep in Huntsville. Galama and Groot did not want to make a decision on their own because the whole question was too sensitive. On the other hand, time was of the essence. The longer they waited the smaller was the chance of finding an afterglow. A quick decision was made; they would wake up van Paradijs. Galama, meanwhile, contacted Richard Strom at the radio telescope in Westerbork so that he could prepare to make radio observations.

'The whole day was spent calling back and forth,' says Groot.'It was a hectic time.' Van Paradijs was in Huntsville, Groot and Galama were in Amsterdam, Costa was in Rome and Heise and Piro were attending a conference in Tokyo along with Chryssa Kouveliotou from the Marshall Space Flight Center, van Paradijs' wife and colleague. Each one had to talk with the other one – and meanwhile the clock was ticking away. Piro remembers vividly how exciting it was in Tokyo. Earlier that day, at about 11:00 Universal Time (only 8 hours after the gamma ray burst), when the X-ray telescopes of BeppoSAX had already been observing the burst position for 4 hours, a rather bright X-ray source was indeed disovered on the edge of the error box of the Wide Field Cameras. 'Right before I was to give my talk in Tokyo, Marco Feroci in Rome sent me the X-ray image', says Piro. Even though it took another three days before it became apparent that the X-ray source was indeed fading, there was no doubt in Piro's mind on February 28th that they were seeing the X-ray afterglow of the gamma ray burst. The news could not have been hotter when he announced to his audience in Japan that BeppoSAX had chalked up a new first.

Costa was now under pressure. If the gamma ray burst showed an X-ray afterglow, then the chances were good that it would also be observable at radio wavelengths and visible light. Naturally, he had alerted Dale Frail in the United States, who would be observing with the Very Large Array at wavelengths of 6 and 20 centimeters. But the radio telescope in Westerbork could also pick up radiation with a wavelength of 36 centimeters, and of course, the more observations that were done, the greater was the chance of success. Above all, a quick decision had to be made. In the Netherlands it was already evening and the constellation Orion would dip beneath the horizon in a couple of hours. If Westerbork didn't get going immediately, the opportunity would not present itself again until Saturday afternoon.

After soul-searching discussions with his colleagues in Rome and Tokyo, Costa came to a decision. Van Paradijs was given the exact position in the sky so that observations with the Westerbork telescope could be made that night. Van Paradijs called Galama, who had already gone home, and Galama contacted Westerbork. Before long, the fourteen dish antennae turned toward the west, and so began the second search for a radio afterglow.

Of course, Galama was wildly enthusiastic. Maybe the Westerbork observations would lead to the first definite identification of a gamma ray burst. It isn't often that a doctoral thesis comes up with such revolutionary results. He was more than willing to give up a night's sleep and a free weekend. He called Groot to tell him the good news. Groot was still at the institute analyzing the optical observations from the

(*Left*) *Paul Groot*'s discovery of the optical afterglow of GRB 970228 revolutionized gamma ray burst research.

(*Right*) *Titus Galama* was the first to discover that there is a connection between gamma ray bursts and supernova explosions.

burst on January 11th because he had scarcely got to it earlier in the day.

'I told Paul that we had the burst position and Westerbork was already working on it,' says Galama. 'Paul realized that on the same night we also had observing time on the William Herschel Telescope.' On La Palma, John Telting would be taking photographs, for the second time, of the position of GRB 970111 in the constellation of the Serpent. The Serpent was visible the whole night but Orion was about to sink below the horizon. Could Telting first take a photo of the position of GRB 970228?

It seemed to be the obvious thing to do but of course, there were a lot of loose ends to tie up. To start with, the observation time on the WHT was granted on the basis of the proposal that was to look only at GRB 970111. In some areas of astronomy, and in particular radio astronomy, it is unthinkable that you just turn around and alter something that was in the original proposal. But luckily, in optical astronomy it is somewhat easier to switch so there was no real problem. Telting could simply look in another direction. But the really critical thing was that Costa had given the burst position to van Paradijs to be used for radio observations with the Westerbork array, *not* for observations with an optical telescope. Nothing had been said about that.

What now? In just over an hour it would not be possible to observe that part of the sky from La Palma, which meant that everything had to be handled with the utmost speed. Galama and Groot called van Paradijs again in Huntsville but he couldn't be reached. They could leave a message on his answering machine but that would mean they would have to wait and that could well take a long time. What should they do? Maybe they should go ahead and let the pictures be made and take care of the political fall-out later. And so it was.

Groot called Telting, who aimed the 4.2-meter telescope, the largest optical telescope in Europe, at a low point, just above the western horizon about halfway between the bright stars Aldebaran and Bellatrix; two images were made. It was now 23:48 Universal Time – less than 21 hours after the actual gamma ray burst.

After taking a catnap, Groot and Galama immersed themselves the following morning in studying the pictures that had come to Amsterdam via the Internet. The William Herschel Telescope has a field of view of about 7 arcminutes, somewhat less than a quarter of the apparent diameter of the full moon. The error box from the Wide Field Cameras on BeppoSAX was a bit smaller (6 arcminutes), but in that little area thousands of stars were visible. One of them could be the optical afterglow they were seeking, but how do you find it? A comparison with the Digitized Sky Survey turned up nothing. But in the Digitized Sky Survey far fewer faint stars showed up than on the WHT photographs.

There was only one thing to do: another picture had to be taken with the William Herschel Telescope so that they could search for that little point of light that had become fainter in the interim. This new photo should be taken, preferably, on Saturday night. Bad luck! La Palma was under heavy cloud cover. On the nights that followed it was still cloudy. It would take some time before anything could be known for certain.

A week passed but finally, on the evening of Saturday March 8th, new photographs were taken at La Palma by Max Pettini from the Institute of Astronomy in Cambridge. To the human eye there was again a sweeping collection of thousands of faint light points, but a computer comparison with the photos from February 28th quickly showed that indeed, one puny little star had disappeared from the stage – a measly little point of light of magnitude 21.3, around one million times fainter than the faintest star that can be seen with the naked eye.[2]

On the photo taken on February 28th, the light source was clearly visible, almost on the edge of the error box of the Wide Field Cameras of BeppoSAX, right next to an even fainter dwarf star. On the pictures taken on the 8th of March the dwarf star could still be seen but there was no trace of the other object. Furthermore, the position neatly coincided with the X-ray source that had been photographed by the LECS detector of BeppoSAX. The X-ray source was observed again on March 3rd and it seemed that indeed it had become fainter, which left no trace of a doubt that this was the X-ray afterglow of the gamma ray burst.

The first optical identification of a gamma ray burst was now a fact. Apparently, the bursts were visible with an ordinary telescope if you happened to be fast enough to catch it. But every minute counts and after a few days there is no point in looking. Thanks to Enrico Costa's Gamma Ray Burst Monitor, thanks to Rieks Jager's and John Heise's Wide Field Cameras, thanks to Jean in 't Zand's software and thanks to the quick action of Paul Groot and Titus Galama, the position of a gamma ray burst could finally be known to within a fraction of an arc second. Now big optical telescopes could be mobilized to search for the source of a gamma ray burst. Perhaps in the place where that little flash

[2] The magnitude scale is a logarithmic brightness scale. The higher the number the less the brightness. The faintest stars that can still be seen with the naked eye have a brightness of magnitude 6. One magnitude difference is approximately equivalent to a factor of 2.5 in brightness; 5 magnitudes difference is a factor of 100.

of light faded into nothingness, there was a very faint neutron star or a very far away galaxy.

A brilliant outcome, but Costa was seething. The Dutch had pulled off a first, in what he thought to be a reprehensible manner because van Paradijs had only been given the burst position to be used specifically and solely with the Westerbork telescope for radio observations. And now his team was publishing one IAU circular after another which made it seem to all the world that he had all the strings in his hands. Nor was this just about the pictures from the William Herschel Telescope. Van Paradijs had also contacted Jorge Melnick, the director of the La Silla Observatory in Chile where ESO, the European Southern Observatory, has a number of telescopes in operation, among which is the advanced 3.5 meter New Technology Telescope (NTT). Melnick was immediately convinced of the importance of the discovery and in the night of the 12th to 13th of March the NTT took photos of the burst position with an exposure time of one hour. This revealed a very faint, stretched-out smudge of light that was visible exactly at the point where the optical afterglow had been seen. The nebular patch was almost certainly a distant galaxy. And although there could naturally be speculation about a chance superposition, it began to seem more and more convincing that the gamma ray burst seen on February 28th had taken place at a distance of many billions of light years.

'Jan is a tough one,' says Costa. 'There is nothing obvious; he is extremely correct and he would never lie, but he is tough. Let's put it this way: he certainly isn't somebody who could read Akkadian. He refused, for example, to synchronize the publication of our *Nature* articles, with the result that the *Nature* article about the optical afterglow appeared before our article about the X-ray afterglow, even though the X-ray one was discovered first.'

The article by van Paradijs and his colleagues was published on April 17, 1997, just seven weeks after the gamma ray burst, and five weeks after the discovery of the faint galaxy. Directly after the identification of the optical afterglow, on March 9th, van Paradijs had contacted the *Nature* editors and in the week after the NTT pictures were taken, the article was written post-haste. To speed things up, *Nature* asked the reviewing referees to react within 24 hours. Never before had so little time expired between a scientific discovery and a publication about it in *Nature.*

'We could never manage that,' says Costa emphatically. 'I am more of a hardware man; Jan is a much better astrophysicist. My poor English probably has something to do with it too. The referees found my article very interesting, but it had to be fundamentally rewritten. That caused a delay of a month.' Costa wanted the two articles to appear in the same issue of *Nature,* but van Paradijs wouldn't hear of it. 'Take it up with the editors,' was his reaction. The *Nature* editors in turn told Costa that it really was a question between him and van Paradijs. The end result was that the article about the X-ray afterglow of GRB 970228 appeared on June 19th, a good two months after the van Paradijs publication.

In May 1997 an international conference was organized on the Italian island of Elba. The remarkable observations of GRB 970228 were presented, but unfortunately the atmosphere reeked of hostility between the opposing groups. 'It was a dreadful mistake,' recalls Groot. 'Nothing good could come of it; they were just hammering away at each other.' Finally, the science steering committee of BeppoSAX came to a decision: the sky positions of the gamma ray busrts as determined by the Wide Field Cameras should, as quickly as possible, be made public through the Internet so that observers all over the world would have the opportunity to track down the afterglows. Costa finally acquiesced but it went against every grain in his body. 'Meanwhile the steering committee had the article by van Paradijs at their disposal, but they didn't have our publication yet,' he says. 'This created the impression that we had not reacted quickly enough and therefore it was better to make the gamma ray burst positions open to everybody.'

It became immediately apparent that making the burst positions freely available would have sensational consequences. Shortly after the discovery of the optical afterglow a number of observing proposals from astronomers streamed into the Space Telescope Science Institute in Baltimore to make use of the Hubble Space Telescope to observe the

Detailed observations were made of the optical afterglows of gamma ray bursts with the Hubble Space Telescope.

fading object. In its orbit around the earth, unhindered by disturbances in the atmosphere, the Hubble Telescope, in spite of the relatively limited diameter of its mirror (2.4 meters) could possibly be helpful in studying the afterglows.

Normally, the observational programs with the space telescope are planned months ahead, so it made no sense in this case to submit a standard proposal. But in anticipation of the eventuality of this kind of situation, where it is impossible to plan ahead, the director of the Space Telescope Science Institute has the flexibility to intervene and make a place for observations where speed is of the essence. All the astronomers concerned called upon this 'director's discretionary time'.

Director Robert Williams didn't quite know what to do with the flood of requests. If he accepted one particular proposal then he would be forced to pass the others by. Rather than do that, he decided to have the observations carried out by a team under the leadership of one of his own people, Kailash Sahu, and then immediately make the results public so that anyone would be free to do his/her own analysis.

On March 26th and April 7th Sahu and his colleagues trained the Hubble Space Telescope on the position of the afterglow and took photographs with an exposure time of many tens of minutes. The photographs were immediately put on the Internet and through an IAU Circular the astronomical community was informed about the availability of the observations. On both days the afterglow could be clearly seen, and indeed, on one side of the pinpoint of light, a hazy patch was discernable, very probably the galaxy in which the gamma ray burst had taken place.

The Hubble observations were published in *Nature* on the 29th of May, simultaneously with an article by Galama and his colleagues about the course of the brightness of the afterglow. In both articles it was hinted that the gamma ray burst seemed to have occurred in a distant galaxy but this could not be said with certainty. It was still possible that a nearby gamma ray burst happened to be observed simultaneously in the same direction as a galaxy billions of light years away.

In the weeks that preceded the publication of both articles in *Nature*, there was enormous confusion about the supposed association of the afterglow with the galaxy. Some astronomers, such as Don Lamb, were so unequivocally convinced that gamma ray bursts originate in the halo of the Milky Way galaxy that they were totally unimpressed by the 'apparent' association with the faint galaxy. After all, it must be borne in mind that since there are countless faint galaxies in the sky, the possibility of a chance alignment is actually a large one.

An Italian research group from the Istituto de Fisica Cosmica (IFC) in Milan, headed by Patrizia Caraveo, had studied Sahu's photographs and came to the conclusion that between the 26th of March and the 7th of April the afterglow had moved very slightly. If the object really was billions of light years away, this would be impossible. The fact that some movement had been measured would indicate, according to Caraveo, that the gamma ray burst occurred much closer.

And as if the confusion were not already great enough, Mark Metzger from the California Institute of Technology in Pasadena, who had made observations with the 10-meter Keck telescope in Hawaii, reported that the nebula-like patch had grown fainter during the month, which would mean that it could not be a distant galaxy at all.

Again it seemed that the panic button had been pushed. The IAU Circulars, sometimes with contradictory information, followed one upon the heels of the other so quickly that nobody seemed to know what was going on. Some astronomers rushed carelessly through their analyses because they were so anxious to publish, and on the basis of skimpy observational results they came out with less than convincing conclusions. Perhaps this should not be surprising: It was 30 years since Vela detected the first gamma ray burst and ambitious astronomers were chafing at the bit to come up with the answer. In many cases these astronomers were in elementary school at the time of Vela and some of them weren't even born yet.

Though it would take until the middle of May before there was compelling confirmation of the cosmological distance scale of gamma ray bursts, the discovery of the optical afterglow of GRB 970228 is generally regarded as the biggest breakthrough in the research so far. The revolutionary observations were the subject of the day during the 4th Huntsville Gamma Ray Burst Symposium in October of that year, and *Science* magazine listed it among the ten most important scientific breakthroughs of 1997. Together with the BeppoSAX team, van Paradijs received the 1998 Bruno Rossi prize from the American Astronomical Society, and in 1999 in his own country he was awarded the physics prize from the Dutch Physical Society.

Ed van den Heuvel, director of the 'Anton Pannekoek' astronomical institute at the University of Amsterdam, says it was no coincidence that van Paradijs and his group succeeded in making the long awaited breakthrough. 'Jan often said that life is a question of priorities. He had a nose for scientific problems that were not only interesting but also held the promise of yielding important results.'

After the success surrounding the February burst, van Paradijs and his group wanted, in the next few months, to set up a worldwide network that would be able to track down afterglows even more quickly in the future. He succeeded in getting many small and medium-sized telescopes to agree to 'override proposals', which meant that proposals from his team had the right to break into an ongoing observational program and take over the telescope if there was a new gamma ray burst to study. He supplied his graduate students with pagers and mobile phones so that at any given moment they could be contacted. In the meantime, he was busy in Amsterdam and Huntsville, firing up the interest and enthusiasm of students for the most hectic domain in all of astronomy.

But it did not appeal to everybody. For Paul Groot the atmosphere was too hurried and chaotic. A year after the February burst, he turned back to his original PhD research on certain types of variable stars: 'a great subject, full of good physics'. But for Titus Galama and his new colleague, Paul Vreeswijk, this excitement was just their cup of tea. They worked at the institute far into the night and never slept without their cell phones under their pillows, and they were never without a list of telephone numbers of telescopes all over the world. After all, you never can tell.

In 1998 Jan van Paradijs was diagnosed with cancer. In a hospital in Huntsville he received chemotherapy, but to no avail. In the summer of 1999 this vital astronomer, weakened by the onslaught of his sickness, flew to Amsterdam, accompanied by his wife, Chryssa Kouveliotou, to enter the hospital of the Free University where he was given the most advanced therapy. Between treatments, when he was not too exhausted, he kept up with the literature, discussed research problems with his graduate students and worked with Chryssa on a comprehensive article on gamma ray bursts.

On October 22nd, while all his colleagues were attending the 5th Huntsville Gamma Ray Burst Symposium, *Science* published his last article that he was to see in print, on the relationship between gamma ray bursts and supernovae. By then it was clear that he had not much longer to live. He didn't make it to the *cum laude* graduations of Paul Groot and Titus Galama on the 7th and 8th of December.

Jan van Paradijs died on November 2nd 1999, at the age of 53, less than a thousand days after that February burst that propelled him to world fame.

7 Eavesdropping on heavenly whispers

'The prettiest are always further!' she said at last, with a sigh at the obstinacy of the rushes in growing so far off as, with flushed cheeks and dripping hair and hands, she scrambled back into her place, and began to arrange her new-found treasures.

At the Eagle Guest Ranch you can be deluded into feeling that you are on the set of a cheap western. The floorboards creak under your feet and at the rough-hewn wooden tables people are drinking colas, beer and whiskey. Mounted heads of game stare at you from the walls in a perpetual state of alarm and old fashioned pistols hang above the open fireplace. A few of the men are wearing baseball caps but most of them have on authentic cowboy hats. And even though they ride in pick-up trucks rather than on horses, their boots still sport pointed spurs. This is Billy the Kid country.

The Eagle Guest Ranch is not only a bar annexed to a restaurant, it is also a supermarket, a gas station and a motel. It is situated at the cross roads where the little town first put down its roots, and is, literally and figuratively, the heart of Datil, a village in New Mexico, a stone's throw away from the Very Large Array. If you ride on about 20 kilometers toward the east in the direction of Socorro, you pass the colossal dish antennae of the radio observatory that give off a ghostly rosy glow in the evening shimmer of the dusk.

There are a total of twenty-seven dishes, each with a diameter of 25 meters, laid out in the form of a giant 'Y'. Each of the three arms of the 'Y' is about 20 kilometers long and the 230 ton dishes can be transported from place to place with the help of a special vehicle that moves them on rails. Sometimes they are side by side in the center of the Y and at other times they are placed so far away from each other that the farthest ones are not even visible.

In the middle of the 1960's, the first plans for the Very Large Array (VLA) were conceived. Radio astronomy was very young; quasars – those far removed galaxies that emit a great amount of radio waves – had just been discovered. But it was not until October 10, 1980, that the VLA was officially put into service by the National Radio Astronomy Observatory (NRAO) in Charlottesville, Virginia.

The local population in New Mexico wasn't too keen on the idea. Anything that came from the government was, by definition, suspect; they coveted their spacious, but closed in, world. New Mexico was already saddled with the secret White Sands Missile Range further to the southeast, and nobody liked the idea of a new installation being set up to look for signals coming from outer space. They all knew that in July 1947 a UFO crashed in Roswell, still closer to the Texas border, and in April 1964 a flying saucer landed in Socorro – secrets enough here in cowboy land.

To ignore is perhaps the most effective form of protest. Every time the farmers and lumberjacks rode off to Datil, Pie Town, Old Horse Springs or Apache Creek, they turned a blind eye to the VLA in every sense of the word. In the Eagle Guest Ranch no one knows precisely what is going on or what the 27 dishes are listening for. One of the local residents is convinced that there is a female alien living in Datil who uses the VLA to make contact with her home world.

Actually, it's not easy to explain how the VLA works. Yes, the dishes receive cosmic radio waves but that doesn't have much to do with listening. If you turned the extremely weak radio waves into sound, you would get a low hissing, rustling and crackling. But for astronomers these hissings and cracklings pack a lot of information. With a radio telescope you can map the distribution of hydrogen gas in the universe, study the structure of galaxies and get on the trail of bizarre objects such as explosive quasars, ejected supernova shells and flickering pulsars.

The signals are very weak and the cosmic radio waves contain remarkably little energy. All the radio telescopes on earth have not collected enough energy in the past fifty years to keep a Christmas tree light burning for longer than a minute. A radio photon has a million times less energy than a photon of visible light. You would need a trillion radio photons to generate as much energy as a single gamma ray photon. X-ray and gamma ray astronomers are used to 'hearing' clanging and screeching, but radio astronomers listen for cosmic whispers.

Twenty-seven dishes are better than one. Not only do you catch many more of the weak radio waves, but by carefully coordinating the signals with the help of a powerful computer, you can look more sharply at what you are receiving. A radio interferometer such as the Very Large Array or the Westerbork Synthesis Radio Telescope can discern many more details than the largest optical telescope in the world. Depending upon the wavelength that is being used for observation and the nature of the object being studied, the dishes must be either at a great distance from each other or very close. They are moved every few months, which is no small undertaking when you realize that it requires a couple of hours to move each one of those monsters.

It is not a coincidence that the VLA was placed on the 2,100-meter-high San Augustin plain in New Mexico. For a telescope with three arms each of which is 20 kilometers long, you need a lot of ground and it has to be flat. You also need a high and dry place to carry out radio observations at short wavelengths because there must be the least possible loss due to absorption in the earth's atmosphere. And it must also be sparsely populated, otherwise the observations will be disturbed by the earth's radio traffic. Finally, the American astronomers wanted the most accessible southerly position in order to observe as much of the southern part of the sky as possible. The extensive, abandoned grassy plain, southeast of Datil, was the ideal spot.

You can well imagine that an eager astronomer working with this unique telescope wouldn't mind being awakened in the middle of the night to study one of the most puzzling phenomena in the universe. Still, Dale Frail did bang the phone down in irritation when Enrico Costa called on January 23rd, 1999, with the message that he had another position available for a powerful gamma ray burst. Frail had just become a father and the nights were hardly restful and never long enough. That new burst was just too much.

Of course this was an exception. If there ever was someone ready to jump at the chance to use the VLA to chase down radio afterglows or gamma ray bursts, it was certainly Frail. When he was called again 15 minutes later that same night by Shri Kulkarni from the California Institute of Technology, he was easily talked into it. Later it turned out that GRB 990123 was one of the most spectacular gamma ray bursts in history (see Chapter 13).

By the beginning of 1999 most high-energy physicists were convinced that radio observations were valuable, but it had not always been the case. Gamma ray bursts were high-energy phenomena that you studied by using gamma ray detectors and X-ray telescopes – after all, it had to do with radiation at wavelengths of less than a tenth of a millionth of a millimeter and photons with energies of a hundred thousand or a million electron volts.[1] Anybody working in the high-energy world knew nothing of the other extreme of the spectrum where waves of centimeters or even decimeters were studied and photons were registered with an energy of a millionth of an electron volt.

It still doesn't sit too well with Frail. The discovery of afterglows at visible wavelengths and X-rays attracted much more attention than the discovery of radio afterglows. But if we are talking about the physical phenomena that exert such a catastrophic force at the place of origin, it was with the radio observations that the most information was obtained. And yes, it has troubled Frail that the Bruno Rossi prize in 1998 was given for the discovery of the X-ray and optical afterglows of gamma ray bursts because it leaves the impression that radio observations are not important.

Frail was born in Toronto, but grew up in Baden-Baden in Germany, where his father was stationed at a Canadian air force base. After high school he went back to Canada where he studied astronomy at the

[1] An electron volt is the amount of energy gained by an electron in passing from a point of low potential to a point one volt higher in potential.

University of Toronto. While he was a PhD student, he had stayed at the VLA for a few weeks but the actual Toronto–VLA transition bordered on the traumatic. Except for a handful of scientists and technicians in the control building, there was not a soul to be seen in that wide and desolate stretch of land. The Eagle Guest Ranch in Datil was the closest restaurant around. The nights were pitch black and dead still; once in a while you could hear the hum of the motors of the VLA dishes that were almost continuously in use. Would he ever get used to this place?

Frail has now lived for years in Socorro and he doesn't want to leave. Compared with Toronto, it's a hole in the wall but then again, when you compare it to Datil, Socorro becomes a metropolis on the Interstate from Albuquerque to El Paso. And here too is the Array Operations Center, housed in a modern building which is built in the form of a Y just like the VLA itself. Frail lives five minutes walking distance from his office and now almost never spends the night on that San Augstin stretch of flatness.

In 1993 he became acquainted with the world of high-energy astrophysics. That summer, together with Kulkarni, he had made a very interesting discovery involving the relation between soft gamma repeaters (see Chapter 12) and the remnants of supernovae. Soft gamma repeaters (SGR's) also produce a tremendous burst of gamma rays, but the radiation is less energetic ('softer') – moreover, one source can produce more explosions with interims of weeks, months, or years. The first soft gamma repeaters to be discovered turned out to coincide with the remnant of a supernova in the Large Magellanic Cloud (the expanding remains of an exploded star). Of course it was possible that this could be a chance alignment but when Frail and Kulkarni found another supernova remnant at the position of the second soft gamma repeater, they felt this was no coincidence. Soft gamma repeaters must be young neutron stars, thought the two astronomers.

Their discovery, which they were quick to publish on September 2, 1993 in *Nature*, didn't go down well with the gamma ray burst community. The research on gamma ray bursts (and on SGRs) had not made much progress and now, here came two outsiders with a potential breakthrough! It's not surprising that Frail was a little nervous in anticipation of giving his presentation at the second Huntsville Gamma Ray Burst Symposium in October of 1993. But anxiety gave way to triumph when the Japanese ASCA satellite observed a new explosion exactly in the middle of the supernova remnant, just a couple of weeks before the symposium.

In Huntsville Frail gave his talk but he also listened carefully to all the other presentations about gamma ray bursts, a subject that was essentially new for him. He heard astronomers tell about their fruitless search for optical counterparts, about determining positions with the help of satellites in the Inter-Planetary Network and about the ideas concerning afterglows. And on the trip back to Socorro, he thought: 'I can do this and I can do it better.'

Opinions differed as to whether or not gamma ray bursts would have radio afterglows. Bohdan Paczyński and James Rhoads had just published their article in which they predicted afterglows. But the British astronomer Martin Rees and his American colleague Peter Mészáros thought the radio waves would be far too weak to be observed on earth. Needless to say, this is not exactly what you tell a person who has the Very Large Array at his disposal. Frail was determined to track down the first radio afterglow of a gamma ray burst.

The first few years were not very rewarding. The positions of the bursts were not accurate enough, which meant that the radio observations would have to scan a large area of the sky. This was best done at low radio frequencies with long wavelengths of around 20 centimeters. Only later did it become apparent that afterglows at that wavelength would not be observable, but nobody knew that in 1993. Another problem was the long wait before getting the IPN positions. Kevin Hurley and colleagues were also looking for optical counterparts and afterglows, which meant they were not in any hurry to circulate their positions and at times they didn't do so at all.

Of course, it is not surprising that Frail was elated over the possibility of working with Costa's BeppoSAX group in Rome in the summer of

1996. Finally, there was a satellite that could give an accurate burst position within 24 hours, at least some of the time. It wouldn't be too long now before the Very Large Array was able to locate a radio afterglow. Besides, his close affiliation with Kulkarni and his group in Pasadena might hopefully lead to the first discovery of an *optical* afterglow, which would then be another American first.

But the going was not easy. At the position of GRB 960720, which became available in the second half of August, nothing was found. Then GBR 970111 was almost a debacle when the detected radio source turned out to actually lie *outside* the error box of BeppoSAX. And when GRB 970228 exploded, Frail and his colleagues were too preoccupied to handle it. 'We were absolutely not ready for this', says Frail. 'Just at that moment, we were busy pulling back our article we had sent to *Nature* – the one about the "supposed" radio afterglow from the January burst.'

As a result, the large telescopes that Caltech had at their disposal were not brought into use to search for the optical afterglow of GRB 970228. And so it came to pass that van Paradijs and his team were the first to behold the coveted phenomenon. Like the burst that occurred in January, the one in February showed no radio afterglow: not with the Very Large Array and not with the Westerbork telescope in Holland. Could it be that the theoretical predictions of Paczyński and Rhoads were wrong?

On Thursday May 8th, 1997, the Americans had better luck. On that day, at 21:42 Universal Time, the Gamma Ray Burst Monitor on BeppoSAX registered a gamma ray burst not longer than 15 seconds and not particularly powerful. It was a typical kind of a burst that often occurs, but in this case the explosion happened to appear in the field of view of one of the two X-ray Wide Field Cameras of the satellite. Within a couple of hours a rough estimate of its position could be determined.

As usual, Enrico Costa called the position through to Frail who instantly turned the twenty-seven dish antennae of the Very Large Array in the right direction – a point in the sky not far from the Pole Star, in the constellation of Camelopardalis. Because the earth's axis points almost exactly in the direction of the Pole Star, it remains at a fixed position, high above the northern horizon. The area close to the Pole Star does show a daily motion, reflecting the earth's rotation around its axis, but it never disappears below the horizon. This 'circumpolar' region is therefore always observable, particularly for a radio telescope that works as well during the day as it does at night.

Frail's first observations were at 01:30 Universal Time, less than four hours after the burst – an absolute record time. At that moment it was early evening in New Mexico; the sun had not yet set. Because the area to be observed was so large (it had a diameter of about 10 arcminutes) observations with the VLA were done again at a wavelength of 20 centimeters. Would it finally work this time?

While making preparations for the radio observations, Frail naturally contacted Shri Kulkarni and his colleagues in Pasadena. As soon as it got dark enough, the little area of sky in question would have to be observed with large optical telescopes. Luck was with him. George Djorgovski, a colleague of Shri Kulkarni, was up on Palomar Mountain at the time and the 5-meter Hale reflector of Caltech was ready to go. Djorgovski was busy with another observation but it would be no trouble for him to get a few images of that piece of sky in Camelopardalis.

'As soon as I got the call from Dale, I knew there wasn't a minute to lose,' says Djorgovski. 'It was just becoming dusk and Europe had a head start on us because it was in the middle of the night there. The images had to be made as quickly as possible, maybe even before it became completely dark.' Although Djorgovski had been born in Europe he had lived and worked in California for 20 years. He felt himself to be totally American and in true American fashion, he went all out to win the race.

Stanislav Djorgovski no longer has any connections with the former Yugoslavia, the land of his birth. He was an only child and his mother and father died when he was a teenager. Stanislav lived for years with his grandparents and as far as his emotional and intellectual development was concerned, he was pretty much on his own. After studying physics and astronomy, which had no strong tradition in Yugoslavia, he left for the United States in 1979. There, he adopted 'George' as his middle name because it was easier for Americans to pronounce.

That evening Djorgovski turned the giant dome of the Hale telescope towards the north and focused the 5-meter telescope on the constellation Camelopardalis. The field of view from the telescope was too small to photograph the large error box of BeppoSAX all at one time, but by making a sort of mosaic of images, it was hoped that he could detect an optical afterglow. Around 9:30 at night the first image was taken.

The Hale telescope was officially put into service in 1948 and was named for George Ellery Hale, who was the driving force in making the telescope a reality. For many years it was the largest optical telescope in the world. Compared with the compact, industrial looking monstrous telescopes of today, such as the Keck telescopes on Hawaii and the European Very Large Telescope in Chile, the Hale reflector is a gracious instrument, smooth-moving in spite of its heavy, classic mounting, housed in an extremely spacious colossal dome, sitting like a jewel atop the shadowy Palomar Mountain.

It would have taken the classic photographic cameras of half a century ago many hours to record untold numbers of faint stars down to 23rd magnitude (almost 10 million times fainter than would be visible to the naked eye) but it was done in a few minutes by the modern electronic CCD cameras of 1997. And one of those faint stars could be the optical afterglow they were all seeking.

In the control room of the Hale telescope, Djorgovski immediately compared the images with the existing ones of the same area in the Digitized Sky Survey, but nothing showed up. If there really was an afterglow, it certainly would be very faint. Djorgovski's colleague, Mark Metzger in Pasadena, could do a more extensive analysis that involved first sending the raw observation data to Caltech by a slow modem connection, but even then Metzger's search also drew a blank. There was just one thing left to do; on the next day the same area had to be imaged to see if, among all those stars, one of them had become fainter.

Meanwhile, van Paradijs' European team was by no means sitting around twiddling their thumbs. On the night of the 8th of May, Titus

The Hale telescope on Mt. Palomar in California has a mirror diameter of five meters and was, for a long time, the largest telescope in the world.

Galama and Paul Groot did not have access to the William Herschel Telescope on La Palma, but thanks to the cooperation of two astronomers from the American Kitt Peak National Observatory in Tuscon, Arizona, images could be made with the 3.5-meter WIYN telescope, a modern telescope under the management of the University of Wisconsin, Indiana University, Yale University and the NOAO (National Optical Astronomy Observatories).

One night later when the WHT on La Palma was available, comparison photographs were made. On Saturday Galama and Groot threw themselves into the job of analyzing the data, which was all the more difficult because the images were taken with different telescopes. In that regard, Caltech had an advantage. Djorgovski not only had a bigger telescope available, he could also observe on both nights with the same apparatus.

However, the optical afterglow of GRB 970508 was not an easy prey to catch. In the error box of BeppoSAX there was no object found that

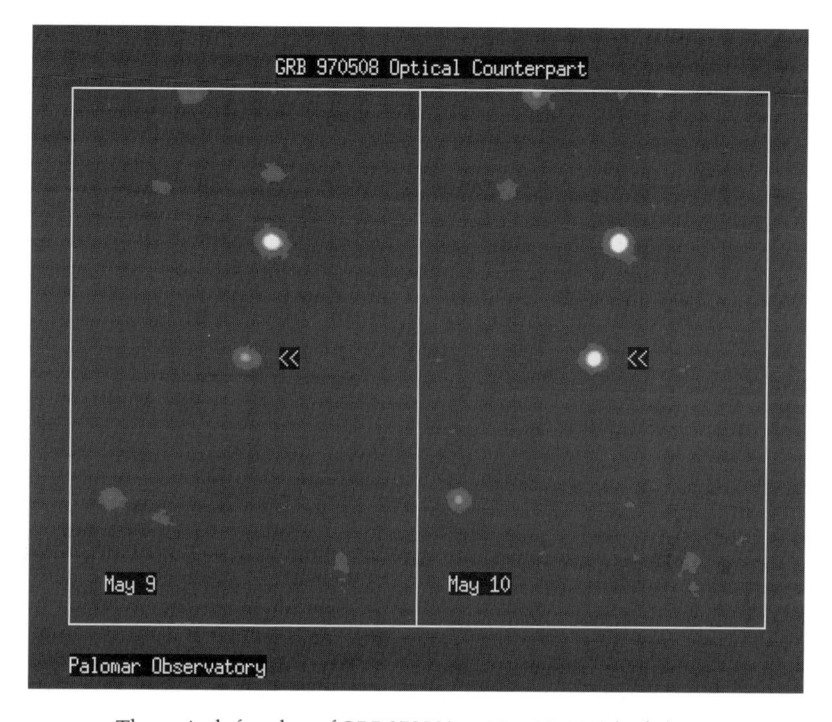

The optical afterglow of GRB 970508 on May 10, 1997 (right) was noticeably brighter than it was on May 9 (left).

faded between May 8th and 9th. Metzger, on the other hand, found a star that had *increased* in luminosity, but of course, that could not be the afterglow they were looking for: the gamma ray burst was long gone wasn't it? It probably was just an ordinary variable star.

But later on, Djorgovski recalls, the feeling of doubt returned. In the meantime BeppoSAX had also discovered an X-ray afterglow which made it easier to fix the sky position of the burst much more accurately. When they took a closer look, the little variable star turned out to be in the smaller error box. Could this perhaps be the optical afterglow after all? Was it possible that the amount of visible light produced by the gamma ray burst would first increase for a little while?

Too much knowledge sometimes leads to too much cautiousness. The Caltech astronomers couldn't imagine that an optical afterglow could increase in luminosity. The Amsterdam team also wanted to do observations again with the William Herschel Telescope on Saturday and Sunday, the 10th and 11th of May, just to make sure. But a relative outsider, Howard Bond, from the Space Telescope Science Institute in Baltimore, didn't hesitate for a moment to publish his observations made with a small 90-centimeter telescope on Kitt Peak on the 8th and 9th of May. He had made images of the gamma ray burst position and on Saturday the 10th he made his discovery of a brightening object known through an electronic circular of the International Astronomical Union (IAU). Later it turned out that this was indeed the optical afterglow that was being sought.

'Bond was ahead of us,' says Djorgovski, 'but we were still the first ones to determine the redshift of an afterglow.' Redshift is the phenomenon in which the light from a very distant heavenly body arrives on earth with a longer (redder) wavelength than it had when it was sent out. As a result of the expansion of the universe, all very distant galaxies display a redshift: their light travels hundreds of millions, or even billions, of years through the expanding cosmos whereby the light waves get stretched out. Once they arrive on the earth they have a longer wavelength than they had at the moment they started out. The longer the light has been traveling (in other words: the greater the distance from the object being studied) the greater the cosmological redshift. As a result, the redshift becomes an unambiguous measure of the distance of a far removed celestial body.

It is not enough to identify a redshift by only looking at the color of the observed object. Astronomers must record the spectrum of the object; a graphic display in which you can see how much radiation is received at each wavelength. In such a spectrum many peaks and troughs can be seen. Some hot gases send out radiation primarily at a small number of very specific wavelengths. At other wavelengths the radiation is absorbed by atoms and molecules that are found between the light source and the observer on earth. For all of these so-called emission and absorption lines, the wavelengths at which they occur are precisely known; and by measuring the amount by which the lines in the spectrum of the heavenly body are shifted, the redshift can very accurately be fixed.

Determining spectra and measuring redshifts are all in a day's work for just about all observational astronomers. But when the object they are looking for is no brighter than 20th magnitude, then they need a very big telescope to measure the redshift. The light received must be broken up into the different wavelengths, but when the star or galaxy is too faint even the most sensitive spectrograph cannot make a spectrum of it.

Caltech is actually a co-owner of the two Keck telescopes on Mauna Kea, Hawaii. With the 10 meter diameter of their mirrors, they are the largest optical telescopes in the world. They catch four times as much light as the 5-meter telescope on Palomar Mountain and they are equipped with such sensitive spectrometers that they can analyze the light from a little star with a magnitude of only 20 or even fainter.

Charles Steidel, a colleague of Djorgovski, Metzger and Kulkarni, was observing distant galaxies with the Keck in Hawaii when Metzger called him with the news about the optical afterglow. On the night of Saturday May 10th to Sunday the 11th, Steidel found time to record the spectrum of the variable object. The rough data were sent to Pasadena over the Internet for Metzger to analyze.

'Nobody could come near the data,' says Djorgovski, 'Mark wanted to do it alone.' There was a lot at stake. From the spectrum it was hoped to determine the redshift which would then make it possible to unequivocally establish the distance of a gamma ray burst for the first time. That had not been possible with the afterglow of February 28th. Although this afterglow coincided with a fuzzy patch that was probably a distant galaxy, astronomers couldn't be absolutely sure that there was an association between the two. Two years after the debate between Bohdan Paczyński and Don Lamb about the distance scale of gamma ray bursts, it might finally be known which one of the two was right. As Djorgovski says, 'you don't get the chance every day to solve one of the important mysteries of the universe.'

On Sunday the 11th of May, 1997, Mother's Day in the U.S., that chance came to Metzger and Djorgovski. 'Mark came into my office with the spectrum in his hand,' relates Djorgovski. 'You're not going to

Hubble Space Telescope image of the optical afterglow of GRB 970508, the first gamma ray burst for which the distance was determined.

believe this,' he began. 'I was ready for anything; Mark even thought he may have found blueshifted iron lines that would indicate that gamma ray bursts are exploding stars in the halo of the Milky Way galaxy. But as soon as I looked at the spectrum, I recognized the absorption line of magnesium at a redshift of about 0.8.'

In the meantime Metzger had identified many more spectral lines. All the wavelengths were 83.5% longer than they should have been. The redshift (referred to by astronomers with the letter '*z*') was 83.5 % (0.835). To put it another way: from the moment the light was absorbed, the universe has expanded by 83.5% That means that the observed light had been under way through the expanding universe for 6 billion years, and by definition the distance is around 6 billion light years.

Strictly speaking, this does not refer to the distance of the optical afterglow itself but to the material that produces the absorption lines in the spectrum. That absorbing material, cool interstellar gas, must of course be somewhere between the afterglow and the earth, which means that the redshift found by Metzger and Djorgovski can only be a measurement of the lower limit of the distance. But since it is quite probable that the absorbing gas is in the same galaxy as the afterglow, its distance, like the galaxy, would also be 6 billion light years.

On that same Sunday, Metzger and Djorgovski sent around an IAU Circular. With carefully chosen, non-committal wording, they wrote: 'If the source is associated with GRB 970508, then it is evident that the gamma ray burst took place at $z > 0.835$'. But no one doubted that the observed little star was indeed the optical afterglow of the gamma ray burst. What was suspected in February was now a certainty. Gamma ray bursts are at cosmological distances and take place in extremely far removed galaxies. The riddle of the distance scale of gamma ray bursts was solved once and for all, thanks to the observations made with BeppoSAX, the pioneering work of Jan van Paradijs and his team, the optical identification by Howard Bond and the Caltech team and the spectral analysis by Kulkarni's group at Pasadena. The breakthrough inspired the theoretician, Ralph Wijers to write the following verse:

> Twinkle, twinkle, gamma star,
> I don't wonder where you are.
> For BeppoSAX pinned down the place
> for Jan's and Howard's optic chase.
> Mark Metzger with his spectrograph
> then wrote your riddle's epitaph:
> $z > 0.835$

The days immediately following the discovery were ecstatic for Kulkarni, Metzger and Djorgovski. Press releases were written and sent out; there were endless telephone calls from colleagues and journalists; they had to convince *Nature* to publish fast (the article appeared on the 26th of June), and at the same time, reports of observations of the ever-diminishing afterglow were pouring in. 'In fact that whole period between the first call from Dale and the publication of the article in *Nature* was a madhouse', declares Djorgovski.

And how was it going with Dale Frail himself? On May the 11th he had still not succeeded in finding a radio afterglow. His first observations on the 8th and 9th of May were done at a wavelength of 21 centimeters because the burst position was not yet accurately known; and the Very Large Array has a relatively large field of view only at longer wavelengths. 'The first attempts were negative', says Frail, 'and when Howard Bond came along with his optical afterglow, it took me a few days to get access to the VLA again.'

Although Frail was a staff astronomer at NRAO, it did not mean that he could use the colossal radio telescope any time he wished to. 'Override proposals as they exist in optical astronomy are unknown in radio astronomy', he says. 'You are always dependent upon the willingness of the observers who are using the telescope at that moment. They can always say, "no". I spend 30% of my time begging on my knees to get access to the telescope.' Unfortunately, most astronomers have little interest in gamma ray bursts, those fleeting phenomena from the other side of the electromagnetic spectrum. Frail was the only one in New Mexico who cared.

On May 13th, 5 days after the gamma ray burst, observations could finally be resumed. Of course they now were done at a shorter wavelength since the burst's position in the sky was already known and a large field of view was no longer necessary. The new measurements at a wavelength of 3.5 centimeters immediately gave off a strong signal. 'When I first saw the radio source, I didn't want to believe it,' says Frail. But 24 hours later the radio source was not only still there but it had become considerably brighter. Moreover, it was also observable at a wavelength of 6 centimeters and very weak at 21 centimeters. You couldn't miss it. After the first X-ray afterglow in July 1996 and the first optical afterglow in February 1997, there was now the first observation of a radio afterglow of a gamma ray burst.

And what an afterglow it was! Very quickly it became apparent that the radio source had enormous fluctuations in luminosity. On average, the luminosity increased steadily but there were great differences from day to day. Sometimes the source could get twice as bright within two hours only to get weaker a short time later. Furthermore, these luminosity fluctuations did not occur at all wavelengths at the same time. It was as if the source varied not only in luminosity but also in 'radio color'. How could this bizarre behavior be accounted for?

The answer came from Jeremy Goodman, who Frail describes as a 'super, super bright guy' from Princeton University. According to Goodman, rather than looking for the cause in the radio source itself, there was talk of scintillation – a kind of cosmic 'twinkling' effect. If you look at the stars in the sky at night you will see some stars that vary a little in brightness (and sometimes in color too). This twinkling is caused by vibrations in the earth's atmosphere; but with the star itself, nothing is happening. The scintillating of the radio sources is due to the same principle but here the effect is not produced in the atmosphere of the earth but in the thin material between the stars in the Milky Way.

Scintillation occurs only when there are so-called point sources. And since stars are really point-shaped light sources they twinkle because of their great distance. On the other hand, although the planets Jupiter and Saturn appear to the naked eye to be point sources, they actually have an apparent size of a few tens of arcseconds, which becomes evident when observed through binoculars or a small telescope. Such extended objects display no scintillation: the luminosity fluctuations at different positions on the planet disk cancel each other out. That is why stars twinkle and planets do not.

Radio scintillation also appears only when there are point sources, but this happens under even more restricted circumstances. The effect disappears as soon as the radio source in the sky is bigger than 3 microarcseconds. That is about as big as the period at the end of this sentence, seen from a distance of 20,000 kilometers. If the luminosity fluctuations of the radio afterglow of GRB 970508 were caused by scintillation, then the radio source must be very small indeed. But that is exactly what you would expect when the radio waves are produced by an expanding cloud from an explosion! It would only be some time after the explosion that the cloud would be so big that, as seen from the earth, it would cover an angle of 3 micro-arcseconds.

Goodman's explanation hit the nail on the head. In the course of the month of May the luminosity fluctuations became less apparent and about a month after the gamma ray burst, they disappeared entirely. Obviously, in four weeks, the explosion cloud had reached an apparent size of 3 micro-arcseconds. Combining this information with the known distance of the gamma ray burst, it could easily be calculated that the explosive fireball had expanded almost with the speed of light (300,000 kilometers per second). And it should be noted that it is at such a speed that the phenomena of Einstein's theory of relativity become important.

Frail couldn't believe his luck. Not only was he the first to discover a radio afterglow, he was also the first one who had something to say about the physical features of gamma ray bursts. Never before were the explosion speeds of gamma ray bursts measured, but the radio observations left no doubt that this had to do with a relativistically expanding fireball. 'To this day it seems that some of my gamma ray friends have trouble appreciating the importance of the radio observations,' he says with regret, referring to the fact that the discovery of afterglows in X-rays and visible wavelengths attracted more attention from the high-energy astrophysicists than the radio observations.

From the control room of the Very Large Array, Dale Frail looks out over the northern arm of the gigantic radio interferometer. The farthest dish antennae are little more that white dots on the desolate landscape. Towards the east, the horizon is dominated by two gently rounded hills that are just as alluring as they were 10 years ago. Much has remained the same as it was back then: the computers that correlate the observations of the twenty-seven radio dishes (once the state of the art but now barely more powerful than a good PC[2]), the impressive silence and the

[2] Over the next few years the VLA will be equipped with new receivers and faster computers. Moreover, eight new radio dishes will be added, increasing the instrument's sensitivity enormously.

isolation of the San Augustin plain where the whispering radio signals coming from the cosmos are constantly monitored, and the indifference of the local population who have never heard of gamma ray bursts and will probably never want to.

But much *has* changed. In 1989 the Compton Gamma Ray Observatory had not yet been launched. BeppoSAX existed only on paper, and Bohdan Paczyński was probably the only one who believed in the cosmological distance scale for gamma ray bursts. Ten years on the question of the source of those mysterious explosions has finally been answered. There is no one who can deny the fact that they are among the most energetic phenomena in the whole universe.

And now that the redshift and the distance of gamma ray bursts were known for the first time, the energy of the explosions could also be calculated. The result: GRB 970508 produced more energy in 15 fleeting seconds than the sun does in 10 billion years. What phenomenon in all of nature has such a totally destructive force on its conscience?

8 Thinking with the speed of light

The knight looked surprised at the question. 'What does it matter where my body happens to be?', he said. 'My mind goes on working all the same. In fact, the more head downwards I am, the more I keep inventing new things.'

Sir Martin John Rees is Astronomer Royal but you would never know it. No ribbons or medals, no haughty remoteness, no putting on of airs. Rees is a small, thin, somewhat bowed man in his late 50s, with a gaunt face, wavy gray hair and friendly, penetrating eyes. He sometimes speaks indistinctly and fast, as if it bothers him that his voice and words cannot keep up with his thoughts. Thinking is his job; the cosmos lives between his ears. Rees is one of the greatest theoretical astrophysicists of the twentieth century.

Martin Rees was born in York, in the north of England in 1942. Fifty years ago the night sky in York was jet black and when it was very clear you could see millions of light years away in the universe. Along with the myriads of stars, star clusters and nebulae, one could even see the fuzzy band of the Milky Way and the vague sweep of light of the Andromeda galaxy.

But young Martin found little interest or excitement in amateur astronomy and he had no telescope in the back garden. For him the fascination lay in thinking about the universe rather than looking at it; he enjoyed contemplating space and time, infinity, order and chaos. As a teenager he devoured the book, *One, Two, Three . . . Infinity*, written and illustrated by the Russian-American physicist, George Gamov, one of the founders of the modern big bang theory. Because learning languages was not his strongest point, Rees chose to study sciences.

One, Two, Three . . . Infinity – from the one came the other. Rees studied math and physics at the University of Cambridge. He did not want to become a mathematician; his thesis would have to have more to do with physics, but what? 'Particle physics was too hard', he says, 'and fluid dynamics was too boring.' Finally, the choice went to astrophysics, 'thanks to my thesis advisor, Dennis Sciama – a fantastic man.' Rees was a classmate of Stephen Hawking, another student of Sciama's. The two great Cambridge brains are still good friends.

'It was in the mid-1960s, a great time for a beginning astrophysicist', he says. The first quasars had just been discovered; the cosmic background radiation, a kind of weak 'echo' from the big bang, had been detected, and cosmology was bounding along at a rapid pace. 'As a young researcher you could immediately get involved and make a relevant contribution, and fortunately, this is still true.' According to Rees, the ratio between the number of problems and the number of scientists in astronomy is 'pleasantly high'.

After a short time at the University of Sussex in Brighton, Rees was named Plumian Professor of Astronomy and Experimental Philosophy at Cambridge University in 1973, a chair he held for 18 years. He was also director of the Institute of Astronomy at Cambridge from 1977 to 1982 and again from 1987 to 1991.

Rees can well remember when the news about the discovery of gamma ray bursts was made known to the world by Ray Klebesadel in 1973. In the 1970s almost everybody believed that gamma ray bursts were caused by explosions on relatively nearby neutron stars. The big question was naturally, what exactly exploded. Rees 'thought about it', as he always describes his work, and came up with a credible scenario. Neutron stars have strong magnetic fields. Whenever a powerful 'star quake' takes place on the surface of a neutron star, it causes disturbances in that magnetic field. It is as if one shakes the field lines and these 'whiplashes' activate strong electric fields in the magnetosphere of the neutron star. That leads to a spontaneous forming of pairs of particles and antiparticles which are accelerated to enormous speeds, which in turn produce high-energy gamma radiation.

'I still believe it makes sense', says Rees. 'Even though we now know that gamma ray bursts occur at great distances in the universe, I believe that such explosions must also exist on neutron stars.' Indeed, it seems that the so-called soft gamma repeaters (see Chapter 12) can be explained by instabilities on the surface of a neutron star. A good idea is never lost.

During the 1980s there was little progress made in gamma ray burst research. Rees was busy with other things like quasars, active nuclei of galaxies, black holes, cosmology and the so-called anthropic principle. This principle states that fundamental properties of the universe appear to be finely tuned to generate complex structures such as galaxies, stars, planets and living organisms. What had begun as a casual interest in the mathematics behind natural phenomena, led naturally to speculations about the boundaries of space and time and about an infinite 'multiverse' of parallel universes. *One, Two Three . . . Infinity.*

But the professorship at Cambridge and the directorship of the Institute of Astronomy demanded so much involvement in administrative obligations that it cut down on Rees's thinking time. In 1991 he gave up both positions with the hope of spending more time on research. At the moment he is research professor in the service of the British Royal Society. In 1992 he was knighted and in 1995 he was named Astronomer Royal.

'In itself, the title doesn't mean a lot', says Rees. 'I get no income or special privileges and there are no obligations.' In the past it was very different. The English astronomer John Flamsteed, who assembled the first detailed star catalogue, was the first to be named Astronomer Royal, in1675, and as such he received payment from Charles II. This was also the year the Royal Greenwich Observatory was founded. 'Astronomy was the first science to be paid for by the government', says Rees. 'There is no such thing as a Physicist Royal because physics was not professionalized until much later.'

For almost 300 years the Astronomer Royal automatically became director of the Royal Greenwich Observatory. In addition to Flamsteed, the title was bestowed upon such famous astronomers as Edmond Halley, James Bradley, Nevil Maskelyne and George Airy. But the protocol did not anticipate that there might be a woman Astronomer Royal. When Margaret Burbidge became director of the observatory in 1971, the dual function was discontinued and Sir Martin Ryle was appointed Astronomer Royal. From that moment on, the position became an honorary title.

After setting aside his roles at Cambridge (where he is still working), Rees was able to devote himself again to the mystery of gamma ray bursts. The first results from the Compton Gamma Ray Observatory confirmed without a doubt that the explosions were distributed uniformly all over the sky. Therefore, they could not come from neutron stars in our Milky Way galaxy; these would clearly be seen to have a concentration toward the plane of the Milky Way. The research world was divided into two camps: some thought that gamma ray bursts take place at a distance of billions of light years away but most believed that they were phenomena occurring in the vast halo of the Milky Way galaxy.

It is true that Rees originally favored the local model, so much so that it almost cost him a hefty bet with Bohdan Paczyński, the foremost protagonist of the cosmological hypothesis. However, Rees feels that he always kept a non-judgmental view on the distance problem. 'I have written theoretical papers about both possibilities', he says. 'I was quite open-minded. That is why I was asked to be the moderator in the debate between Lamb and Paczyński in April 1995.'

The occasion is still fresh in his memory. 'Too bad that those present were not asked to vote before as well as after the debate', he says. 'Don Lamb had, by far, the best story and the best arguments. I am quite sure that he convinced many of the people.' It ultimately turned out that Lamb was wrong, but according to Rees, this is what makes the sociological side of the practice of science so interesting: it is when you really believe in something with all your heart and soul that you give your best presentations.

When Jan van Paradijs and his colleagues found the first optical afterglow of a gamma ray burst in 1997 and discovered that it coincided with a dim, distant galaxy, Rees could see that Lamb was not right. And when the redshift of GRB 970508 was measured a few months later, it became irrefutable. Gamma ray bursts occur at a distance of billions of light years away. 'From that moment on I stopped looking at the halo model,' says Rees.

The theoretical problem, of course, did not become easier. If gamma ray bursts are at cosmological distances, they must produce an unimaginable amount of energy and this energy is released in a very short time; some bursts reach their greatest gamma luminosity in a small fraction of a second. From the speed at which the burst reaches its peak brightness, you can figure how big, or rather, how small the area is in which the explosion takes place. An explosion that is stretched out over hundreds of millions of kilometers can never display a luminosity peak lasting less than a second. As a result of the finite speed of light (300,000 kilometers per second) the radiation from different parts of the explosion reach the earth at different moments and so the luminosity peak will always be much broader.

Instead, the peak brightness of a gamma ray burst is often reached in less than a thousandth of a second. That means the size of the explosion, at that moment, can never be greater than a few hundred kilometers – the distance that a beam of light travels in a thousandth of a second. All the energy that is radiated in that first fraction of a second must come from an amazingly small volume, which, in turn, implies that there must be an incredible amount of matter present in that small volume. Even if nature were able to convert all matter into energy fully (this would be according to Einstein's famous formula of $E=mc^2$), there would still need to be a tremendous amount of material to explain the observed energies.

And that is where the problem lies. Hot gas with a high density is not transparent. Our own sun is a good example. The radiation we get from the sun comes from the surface and not (at least, not directly) from the inside where the energy is created. Radiation from the inside cannot easily get out because the sun is not transparent. So if there were an enormous amount of energy suddenly released in the inside of the sun, it could not be radiated outwardly. Instead, a fierce radiation pressure would build up which would blow the sun apart and convert the radiation energy, for the most part, into kinetic (motion) energy.

But how is it then possible that a gamma ray burst can radiate so much energy in so short a time? Judging from the speed with which the burst reaches its maximum luminosity, it must follow that the explosion takes place in a small volume, and judging from the prodigious energy produced, there must be a lot of material in that volume. Yet, in some way or another, an enormous amount of radiation is able to escape.

Rees realized that the observed properties of gamma ray bursts could never be explained away as just an 'ordinary' explosion. He had been faced with a similar sort of problem before when he was studying quasars, the explosive nuclei of distant galaxies. There too, so much radiation comes free in so short a time and in so small a volume that it cannot in any way be considered a regular explosion. Apparently, we are dealing here with a rather universal problem, according to Rees.

The solution to the problem, he thought, lay in the model of the relativistic fire ball that he had calculated and wrote about for the first time in 1966 while he was still a student. The term 'relativistic' is used by physicists and astronomers to refer to the phenomena and processes that take place under such extreme circumstances that they show the effects of Einstein's theory of relativity. With the help of Einstein, Rees was able to tame the gamma ray bursts.

The theory of relativity – there are actually two, the special and the general – is one of the most important theoretical constructs of the twentieth century. The theory of special relativity dates from 1905 and describes the behavior of objects which move with a constant speed with respect to each other. The 1915 theory of general relativity also describes accelerations and in fact, forms a theory of gravity that is much more accurate than that of Isaac Newton.

The theory of relativity is not easy to understand, nor to explain in simple terms. The British astrophysicist Arthur Eddington, who made Einstein's theory known to the world, was once asked if it was true that there were only three people in the whole world who really understood the theory. At which point Eddington stopped to think for a moment,

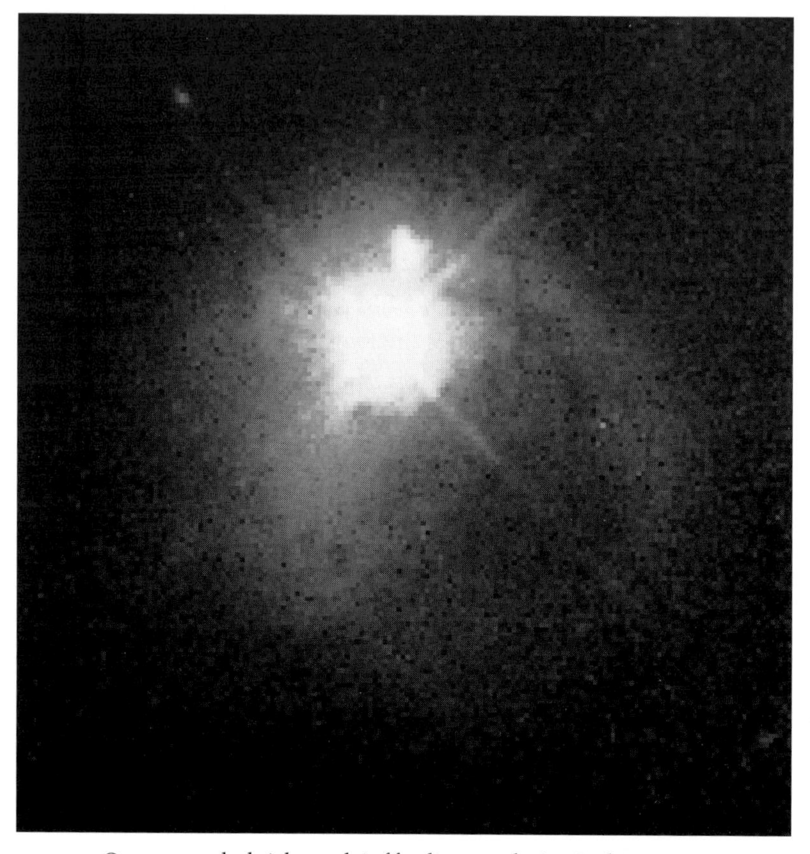

Quasars are the bright nuclei of far distant galaxies. In the enormous energy produced by quasars, relativistic effects also play a role.

then asked, 'And who is the third?' The reason the theory is so hard to fathom, is due to the fact that in our everyday life we never have to deal with the things the theory describes. They happen only in extreme situations where there are phenomenal speeds, superstrong gravity fields, or both.

One of the extaordinary consequences of special relativity is that the speed of light (about 300,000 kilometers per second) is the absolute maximum speed obtainable in nature. Nothing and no one can move faster than light. There is no use trying to control the obedience to this cosmic speed limit because it is physically impossible to violate. Moreover, according to the theory, the speed of light is a true fundamental constant in nature and is independent of the movement of either the light source or the observer.

In our daily lives we are accustomed to increasing or decreasing velocities. For example, a policeman has a pistol that can fire a bullet at a speed of 500 meters per second. But if he is shooting from a moving car, the bullet has a different speed relative to the ground. If the police car is moving at 180 kilometers an hour (50 meters per second) and he shoots in a forward direction, then the bullet has a speed of 500 + 50 = 550 meters per second. But if he shoots to the rear from the moving car, then the bullet, in relation to the ground, is moving at 500 − 50 = 450 meters per second.

But at very high speeds this simple arithmetical solution no longer applies. Once we are dealing with speeds that are a significant fraction of the speed of light, we must use other formulas. If the bullet has a speed of 270,000 kilometers per second (90% of the speed of light) and the police car is traveling at 150,000 kilometers per second (50% of the speed of light), then the resulting speed is not 270,000 + 150,000 = 420,000 kilometers per second, but 'only' about 290,000 kilometers per second, 96.5 % of the speed of light![1]

The same relativity formulas show that a light beam must always move with the speed of light. Even when the light source moves at extremely high speeds, the photons that are sent out have a speed of 300,000 kilometers per second in relation to the observer. It is this unalterable quality of the speed of light in the universe that plays such an important role in Rees' relativistic fireball model.

Suppose an explosion, occurring at some great distance in the universe, packs such a force that matter is ejected into space with almost the speed of light. The actual cause of the explosion isn't even that important – it might have been a collision of two compact neutron stars

[1] Einstein's formula for adding up the speeds which approach the speed of light is $u=(u'+v)/(1+u'v/c^2)$, where u is the sum of the speeds u' and v, and c is the speed of light.

or the catastrophic death throes of a massive star. The critical point is that when huge amounts of matter are being blown away at speeds up to 99% of the speed of light, relativistic effects come into play.

In such a relativistically expanding fireball, high-energy radiation can be created in several ways. To begin with, shock waves occur within the fireball for the simple reason that not all the material that is blown away travels precisely at the same speed. At places where the 'slower' material is caught up by somewhat faster moving material, it is compressed and the gas that is hurled away becomes even more heated and ends up sending out powerful gamma rays. It is these internal shock waves that could be responsible for the initial gamma ray bursts.

After a short time the expanding fireball also interacts with the thin material in the space between the stars. The cloud from the explosion acts as a kind of snow plow and pushes out the interstellar gas and material at high speed. This in turn forms another shock wave which gets terrifically hot and leads to the creation of X-rays, visible light and radio waves. The end result of this frenzied progression is a gamma ray burst afterglow.

Finally, there is the possibility of a 'reverse' shock wave in the direction of the center of the explosion. In fact, this has to do with the 'reflection' of the 'forward' shock wave that comes from the interaction with the interstellar material. All these shock waves produce enormous amounts of radiation at a variety of wavelengths. In most cases this is so-called synchrotron radiation, which is produced when electrically charged particles move at high speeds in an extremely strong magnetic field.

However, the particles that send out this very energetic radiation come to us almost as quickly as the radiation itself. This radiation always travels at 300,000 kilometers per second (the invariable speed of light) but the material moves perhaps 1% slower. The observer on earth then sees a very short burst, even though the radiation that is being observed was sent out over a much longer period.

An example might make this clearer. Imagine that a police pistol is

Martin Rees (*left*) and *Peter Mészáros* joined forces to come up with a theoretical explanation for gamma ray bursts.

so designed that the fired bullets always have a speed of 500 meters per second relative to the ground, regardless of the speed of the car out of which they are fired. The bullets represent photons, the 'absolute' speed of 500 meters per second matches the constant speed of light. The police car rides on a straight road from A to B with A being the center of the explosion and B the observer on earth. The police car plays the role of the relativistically expanding fireball; it has a speed of 495 meters per second, 99% of the 'speed of light'.

We now ask the policeman to fire a bullet every 10 seconds for one minute in the direction of B. In total, seven shots are fired. There were 60 seconds between the first and last shot, but the observer at B sees the seven bullets fly by in a time span of only 0.6 seconds! How can that be? The answer is easy. Bullet number seven is fired 60 seconds after bullet number 1. The speed of the police car is 5 meters per second less than the speed of the bullets, therefore, in 60 seconds the police car has fallen $60 \times 5 = 300$ meters behind the bullet that was shot first. So bullet number 7 is just 300 meters behind bullet number 1, and because the bullets have a speed of 500 meters per second, the observer who is standing still sees them pass in 0.6 seconds. It is as though time has been compressed by a factor of 100.

This explanation accounts for the extremely rapid increase in the luminosity of gamma ray bursts. Because the fireball expands at 99% of the speed of light and the radiating material comes towards us almost as fast as the radiation itself was sent out, a hundredfold 'time compression' occurs. On the earth we see how a burst reaches its maximum brightness in a fraction of a second, but in reality the explosion lasts much longer. Anyone who does not take these relativistic effects into consideration arrives at the wrong conclusions for the dimensions, the density and the energy production of the explosion.

During the 1990s Rees, together with Peter Mészáros of Pennsylvania State University, wrote a number of convincing articles about the relativistic fireball model, to explain the observed properties of gamma ray bursts. The original problem – how can so much radiation escape out of a very small volume in so short a time – was effectively solved by the model. Yes, it is true that at first most of the energy comes free in the form of motion, but through the shock waves this kinetic energy is quickly converted into radiation. That happens in a much larger area with a much lower specific density. As a result of the relativistic effects it seems as though all this radiation is actually sent out in a very short time and therefore comes from a very small volume.

Not that all theoretical problems can be explained away easily. 'For example, there is not yet much known about the role magnetic energy plays in gamma ray bursts', says Rees. 'We would also like to know how much material the relativistic fireball actually contains and to what extent a gamma ray burst is influenced by its environment.' Unfortunately, this is a purely theoretical exercise, he says, because laboratory experiments are not possible.

Thanks to the work of Rees and Mészáros, the relativistic fireball model is now generally accepted as the standard model to explain gamma ray bursts. The radio observations of GRB 970508, which showed indeed that there had to be an explosion cloud that expands almost with the speed of light, constituted convincing evidence for a theory that nobody really doubted. Whatever the cause of the actual explosion, the gamma ray burst occurs because large amounts of matter are hurled into the universe at 99% of the speed of light. Without being able to call upon Einstein's theory of relativity, the properties of gamma ray bursts cannot be explained.

Incidentally, the relativistic fireball model was confirmed not only by the radio observations of GRB 970508, but also by the behavior of the afterglow of GRB 970228. The X-ray afterglow, like that of the optical afterglow, decreases in luminosity in accordance with a so-called power law, whereby the luminosity abates quickly in the beginning and thereafter dims more and more slowly, comparable to the way the reverberations from the striking of a gong fade away.

Ralph Wijers, a theoretician at the State University of New York at Stony Brook, can vividly recall the excitement when the fireball theory was confirmed. Wijers was a postdoc at Cambridge in 1997 and worked closely with Rees and Mészáros at that time. 'When the existence of the optical afterglow was confirmed on March 8th, we immediately began to write an article that was finished within a week and was accepted in three days for publication in the *Monthly Notices of the Royal Astronomical Society*', he says.

For Wijers, who did his thesis on X-ray double stars with Ed van den Heuvel and Jan van Paradijs in Amsterdam, this was not his first meeting with gamma ray bursts. In the beginning of the 1990s he worked for a year at Princeton University where Bohdan Pacyzński was racking his brain over the distance scale of gamma ray bursts. 'Bohdan and I were usually at work at 8 o'clock in the morning,' says Wijers. 'His office was two doors further up and for half a year we discussed gamma ray bursts every morning. I took care of the coffee, Bohdan supplied the cookies.'

Ralph Wijers, together with Titus Galama, deduced the physical properties of a gamma ray burst from the observed spectrum.

But after a while Wijers began to find the subject frustrating. The observational material remained discouragingly scant and most of the theories were based upon statistical arguments. 'As a theoretician you can talk until you're blue in the face,' says Wijers, 'but finally you just need to have hard data. Take the distance scale; on the basis of statistics from 2,000 gamma ray bursts no one was convinced that Paczyński was right, but with the observation of just one redshift in 1997, it was enough to make it clear that gamma ray bursts are at cosmological distances.'

In 1994 Wijers went to Cambridge for four years and divided his study between gamma ray bursts and X-ray double stars; at the end of 1996 gamma ray bursts fell by the wayside. 'Until shortly after that, the first afterglow was discovered', he says. At last there were real measurements available and one could seriously start working on calculations that focused on the different theoretical models. Now Wijers spends at least 90% of his time on gamma ray bursts, concentrating on theories about afterglows.

However, one thing has become quite clear: there is not one gamma ray burst that sticks exactly to the standard model. No two afterglows are alike and the progression of the luminosity often shows deviations of tens of percents. 'The observations are, unfortunately, not yet accurate enough to make distinctions between the various possible explanations for those deviations', says Wijers, 'but we hope that will change in the near future.'

Wijers's favorite burst is still the one on May 8th, 1997; the first gamma ray burst for which a redshift and a distance were determined. Thanks to the organizational talents of Jan van Paradijs and the exertions of Titus Galama, the afterglow from this burst was observed on just about every possible wavelength, from radio waves and infrared radiation to visible light and X-rays. The resulting spectrum (a graphic display in which the luminosity is plotted as a function of the wavelength) was jokingly called the UN graph because observations were made in so many countries.

The shape of the spectrum can be described by five characteristic numbers, explains Wijers, all of which, in the case of GRB 970508, can be measured. And these five numbers are determined by five unknown physical quantities, such as the total amount of energy in the gamma ray burst and the density of the surrounding medium.

Wijers and Galama realized that the five physical quantities – the physical properties of the gamma ray bursts – could, in principle, be calculated from the five numbers that characterize the spectrum of the burst. This is no easy task since each physical property has, in a specific way, an influence on each of the five spectral characteristics.

'The formulas in our article were so long that they didn't fit the two-column format of *The Astrophysical Journal*,' says Wijers, beaming with pleasure. 'The article had to be spread over the entire width of the page.' Five equations with five unknowns is an algebraic problem math students have in high school and they know that it has an unambiguous answer. Thus, the total amount of energy could be determined as 3×10^{52} ergs (over twenty times as much as the sun sends out in ten billion years),[2] and the density of the surrounding interstellar material was determined to be 30,000 particles per cubic meter. Not that the

[2] The so-called scientific notation used to write down large numbers makes use of the powers of 10. For example, 10^{52} (ten to the fifty-secondth) means a 1 followed by 52 zeros.

results are very reliable (the data are not accurate enough for that), but Wijers and Galama had shown that, in principle, it is possible to deduce the physical properties of gamma ray bursts from the observations of afterglows.

Wijers is convinced that the theories can be enormously refined. 'The different models make very definitive statements about the observable properties of the afterglows, such as the relation between the amount of energy in the X-rays and in visible light', he says. 'When more and better observations are available, it should be possible to rule out some of the theories.'

In particular, there is very little known about the very early phases. Most of the time, the first observations of afterglows are made only a few hours after the actual gamma explosion has taken place. 'That is the disadvantage of the power law,' says Wijers. 'In the beginning the luminosity decreases so rapidly, you have to run yourself ragged to follow it. But fortunately, the brightness fades more slowly later and if you can once get to the afterglow, you can study it for a good while.'

No one seems to doubt the validity of the relativistic fireball model, but Sir Martin Rees remains down to earth about its success. 'Someone else can always come along with a better theory', he says. 'In a few years, when we have redshifts and identifications of several hundred gamma ray bursts so that we can seriously look for statistical correlations among the various properties, in principle it could happen that we are looking at it all wrong.' But that is the lot of the theoretician. Every phenomenon ultimately has only one explanation, so most of the theories and models that have been designed over time must, of necessity, be wrong. However, the chance that the relativistic fireball model will ever have to make way for a completely different theory seems rather small, simply because the pieces of the puzzle fit together so well. Meanwhile, Rees has been engrossed for some time in the question of what actually causes the cosmic super explosions. Is it the catastrophic destruction of a massive star, or the equally devastating collision between two compact neutron stars? 'Perhaps both occur', he says. Indeed, the thousands of gamma ray bursts that have been observed by the Compton observatory fall clearly into two groups: 'short' bursts that last less than a second and send out a relatively large amount of high-energy gamma rays, and 'long' bursts that last for a couple of seconds to a few minutes and produce relatively 'softer' gamma rays. 'The short bursts might well be caused by merging neutron stars', according to Rees.

Theoreticians are a long way from being finished with gamma ray bursts, but they don't have a lot of room to play around with. From the observations it is apparent that a gigantic amount of energy comes free from a gamma ray burst. There are very few conceivable phenomena that can produce so much energy. Nor can theoreticians let their unrestrained fantasies seriously conjure up a science fiction scenario because any viable theory must obey the limitations demanded by the laws of nature. So there they sit, caught between the flood of observations and the inflexible wall of nature's laws. The landscape is inspiring but there isn't much elbow room – as became evident in December of 1997. For what do you do when you are caught between a rock and a hard place?

9 Competition for the big bang

'Now tell me, Pat, what's that in the window?'
'Sure, it's an arm, yer honor!' (He pronounced it, 'arrum'.)
'An arm, you goose! Who ever saw one that size? Why it fills the whole window!'
'Sure, it does, yer honor: but it's an arm for all that.'

'Think big' is the motto of Shrinivas Kulkarni. Do you want to get something? Then go for it. Don't settle for half an answer. Use the biggest telescopes in the world and pull in the best people. Do it right or don't do it at all. Think big. Appropriately, he has in the corner of his office on the campus of the California Institute of Technology (Caltech) in Pasadena, a huge pencil, almost 2 meters long.

Kulkarni himself is a small man with black hair, a black mustache and lively dark eyes behind big glasses. But being big is something else again than thinking big. Small people can get big results when they have access to the best instruments: the 10-meter Keck telescope in Hawaii, for example, and the Very Large Array in New Mexico, the Hubble Space Telescope, the Chandra X-ray Observatory. Don't be modest. It's all or nothing.

Kulkarni was born in India and came to the University of California in Berkeley in the United States in 1978, where he was a classmate of George Djorgovski. At Caltech he feels at home. *Big thinking* fits in there. Caltech's 5-meter telescope on Mt. Palomar was the biggest optical telescope in the world for many years. 'Make no mean plans,' said George Ellery Hale, the man behind the building of the telescope. Mediocre ideas get you no place. Need a bigger telescope? Together with the University of California, Caltech built the 10-meter Keck telescope in Hawaii. Still not satisfied? In the middle of the 1990s they built a second one. The design for a 30-meter telescope is ready to go.

(*Left*) *Dail Frail* was the first to discover radio waves from a gamma ray burst.

(*Right*) *Shrinivas Kulkarni* is leader of the gamma ray burst team at the California Institute of Technology.

Big thinking also means making good choices. You can't be the best in everything. Selection is essential, and it is the master who knows how to limit. Which subject offers new perspectives? In what areas can one expect to make breakthroughs? Where do we really find something that needs to be discovered? Make that choice, get immersed in the subject and give yourself over to it unconditionally; because thinking big also means hard work.

Kulkarni used to begin his day at 4:00 AM. That doesn't always work now, but at 6 AM he is there. No place is as dear to him as his office. His wife and two children are used to it. His total dedication to his work was part of him before he got married. But indeed, the past years have been particularly hectic. Gamma ray bursts have their own agenda with no respect for earthly time zones or biological clocks.

Kulkarni has been working on gamma ray bursts since 1993, incredibly long for him. In the past he never stayed with one subject more than 5 years. If you haven't gotten significant results in 5 years then it's time to look for something else. If you succeed in getting somewhere with the subject in 5 years then you have to be careful that you don't let yourself get in a rut. It's time to change; time to blaze new trails.

In the beginning of the 1980s he, along with Dale Frail and Don Backer, discovered the first millisecond pulsar, a supercompact star that turns 642 times per second on its axis and is still the fastest rotating heavenly body that has yet been observed. Later, he directed his thoughts to the structure and the properties of the tenuous material between the stars, not only in our Milky Way galaxy but also in other galaxies. He developed interferometers for infrared radiation and visible light. These are instruments that make observations at other wavelengths as sharp as those that radio astronomers are accustomed to. In 1995, he proved the existence of brown dwarfs – 'failed' stars, no bigger than the planet Jupiter in which no hydrogen fusion takes place. Now, for the past 7 years it has been gamma ray bursts, and it is too exciting to leave.

In the beginning Kulkarni had his doubts. In 1993, when Dale Frail proposed that they set up an observational program for gamma ray bursts, his first reaction was that everybody is doing that already. It's not easy when you always strive to be the first and the best. But it was soon apparent that the high-energy physicists really had no experience with radio observations. In that area Frail and Kulkarni had a head start on them. If only the Very Large Array convincingly determined an accurate position just once, photographs could be taken with the Hale telescope, spectra could be made with the Keck and that could mean the Americans would probably be the first to discover the optical afterglow of a gamma ray burst and the first to determine its distance.

The bursts on January 11th and February 28th 1997 were disappointing, but the radio observations and the fix on the redshift of GRB 970508 made up for that. Optical afterglows turned out to be so very faint that it was only with a giant telescope like the Keck that a detailed

The distance of GRB 971214 was determined with the 10-meter Keck telescope on Mauna Kea, Hawaii.

spectrum of such a weak object could be obtained; they were able to pin down the redshift and the distance. And as co-owners of the Keck, Caltech had a monopoly on the telescope.

The Keck is the biggest optical telescope in the world. Its 10-meter mirror is not made out of one piece of glass but rather out of numerous hexagonal segments. The precise position of each segment is meticulously controlled by a computer and at the same time the curved shape of the mirror surface is continually monitored and adjusted as needed. With its enormous light-gathering power, the Keck is pre-eminently suited for spectrographic observations of extremely faint objects.

The Keck was named for the oil magnate who supplied most of the funds for building the telescope. It came into service in late 1991 on top of the 4200-meter-high volcano Mauna Kea in Hawaii, and there is now a second identical telescope on the mountain 100 meters away from the first. The two giant 'eyes' work independently, but they can also be used as an interferometer, just like the individual dish antennae of the Very Large Array. By aiming them towards the same heavenly body and by carefully combining their observations, they will be able to produce pictures as sharp as an imaginary telescope with a diameter of 100 meters.

Mauna Kea is an astronomical paradise. Besides the two Keck telescopes there are a number of other instruments, among them, the international Gemini North telescope, with a mirror diameter of 8.1 meters, and the Japanese 8.3-meter Subaru telescope. At a height of more than 4 kilometers above sea level the astronomers have little trouble from vibrations in the earth's atmosphere; the top of Mauna Kea is almost always above the clouds and at such an elevation the observatory is also wonderfully suited for infrared and submillimeter wavelength observations.

Because of the altitude, however, it is no fun staying up there for a prolonged length of time. The air is thin and the lack of oxygen leads to headaches, tiredness and loss of concentration. Astronomers and technicians who must remain for a while first get acclimatized by spending a night at the guest house at Hale Pohaku halfway up the side of the gigantic volcano. And whenever it is possible, the observations with the Keck are done by remote control from the control center in Waimea at the foot of the mountain, or even from the mainland, via a fast Internet 2 connection.

On Sunday, December 14th, 1997 the white domes of the two Keck telescopes were still basking in the sun when, at 23:21 Universal Time, there was a strong gamma ray burst in the constellation of Ursa Major. The burst lasted more than half a minute and was observed by the BATSE detectors on board the Compton Gamma Ray Observatory and by the Gamma Ray Burst Monitor of BeppoSAX.

On BeppoSAX, a gamma ray burst hadn't been in the field of view of one of the X-ray Wide Field Cameras since the 8th of May. But the script was ready. Less than 7 hours after the explosion, the LECS and MECS X-ray telescopes of the satellite were aimed at the position of the explosion and the Italian team discovered a new source of X-rays for which it was possible to fix the position in the sky to within an arcsecond.

That same night there were a number of telescopes in the United

Plates

NASA
TRW

With the help of the robot arm of an American space shuttle, the Compton Gamma Ray Observatory was placed into orbit around the earth in April 1991.

(LEFT) The Compton Gamma Ray Observatory in one of the test rooms of TWR Inc. The Compton is the heaviest scientific satellite that has ever been launched.

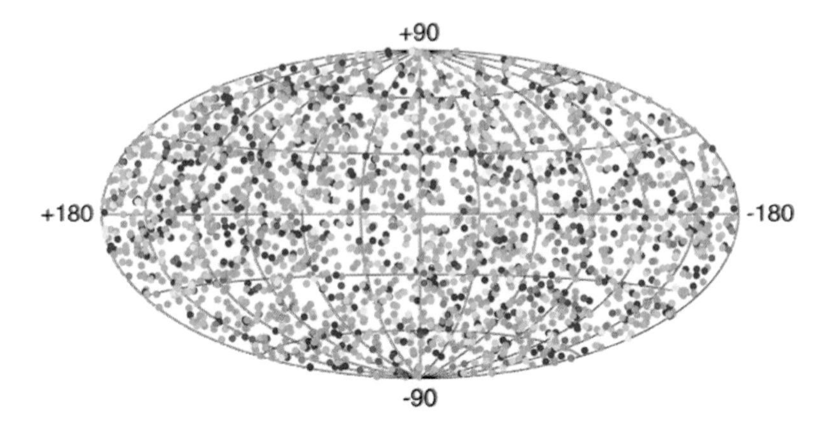

By the beginning of 2000 BATSE had detected at least 2700 gamma ray bursts. They came from every possible direction, as can be seen on this map of the whole sky.

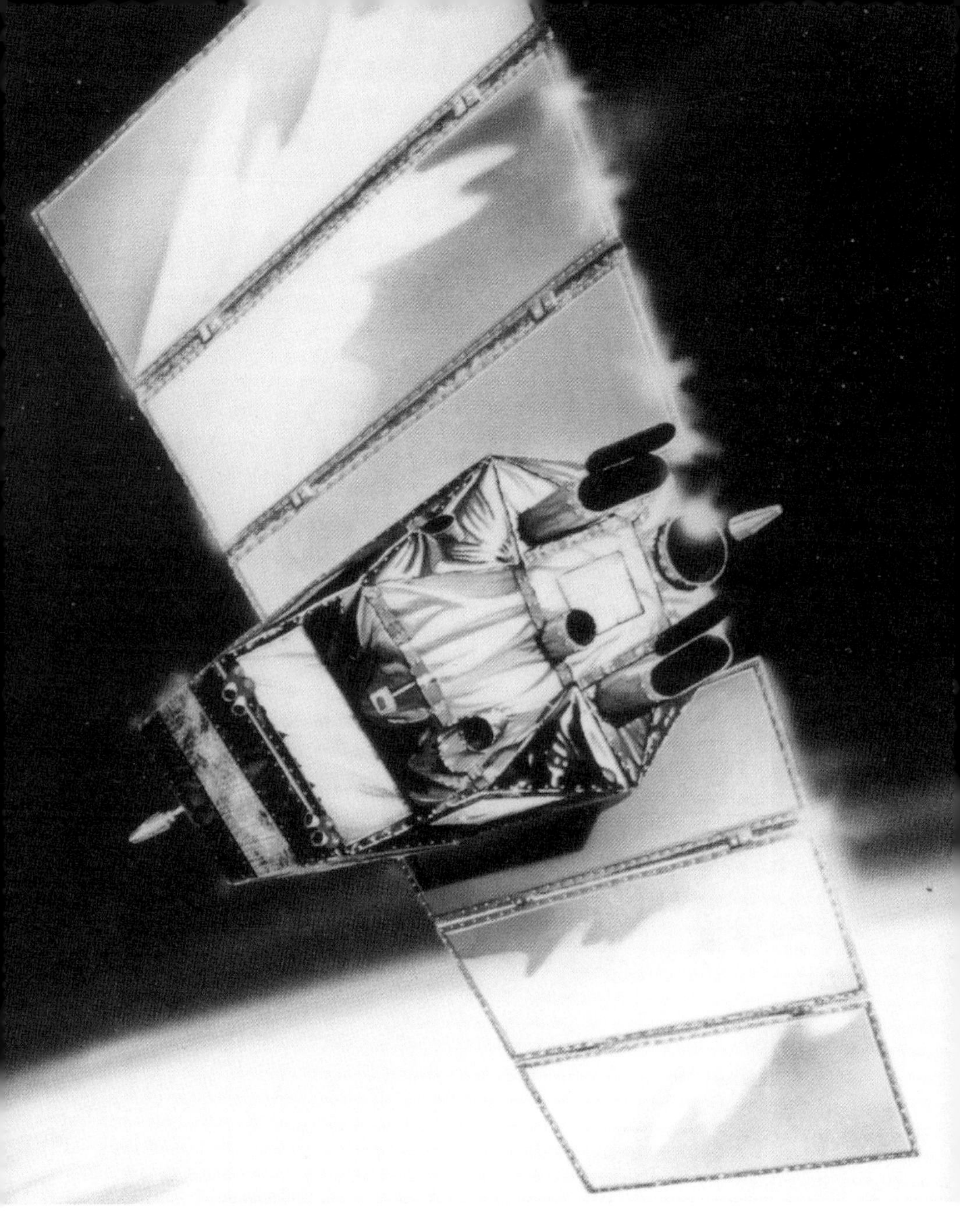

An artist's impression of BeppoSAX in orbit around the earth.

The Pegasus rocket that placed HETE in orbit around the earth was launched from the underside of an airplane wing.

On board the Russian space station Mi was the Utrecht COSOS camera, one of the predecessors of the Wide Field Cameras on BeppoSAX.

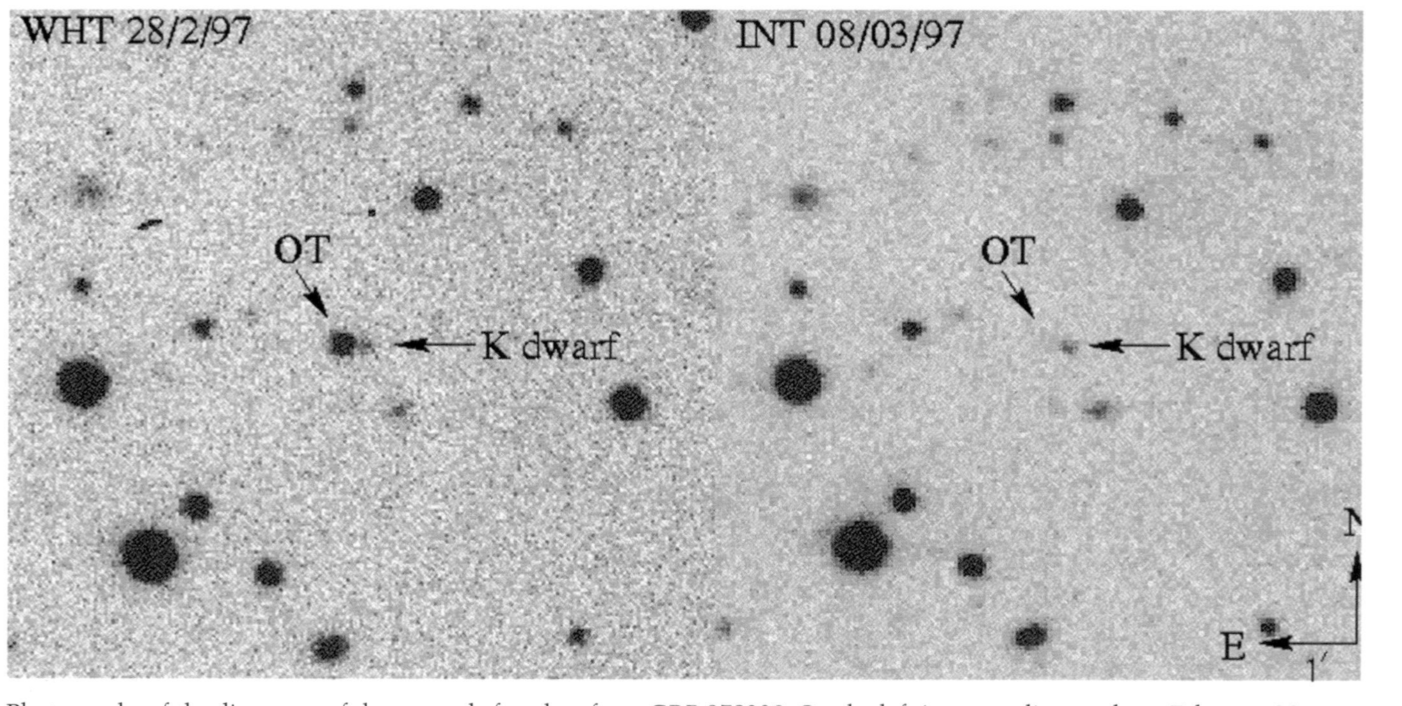

Photographs of the discovery of the optocal afterglow from GRB 970228. On the left is a recording made on February 28, 1997 with the William Herschel Telescpe. Near to a weak dwarf star ('K dwarf'), a rather clear afterglow can be seen (OT stands for optical transient) Right: a recording made on March 8, 1997 with the Isaac Newton Telescope. The dwarf star is still visible; the afterglow has faded away.

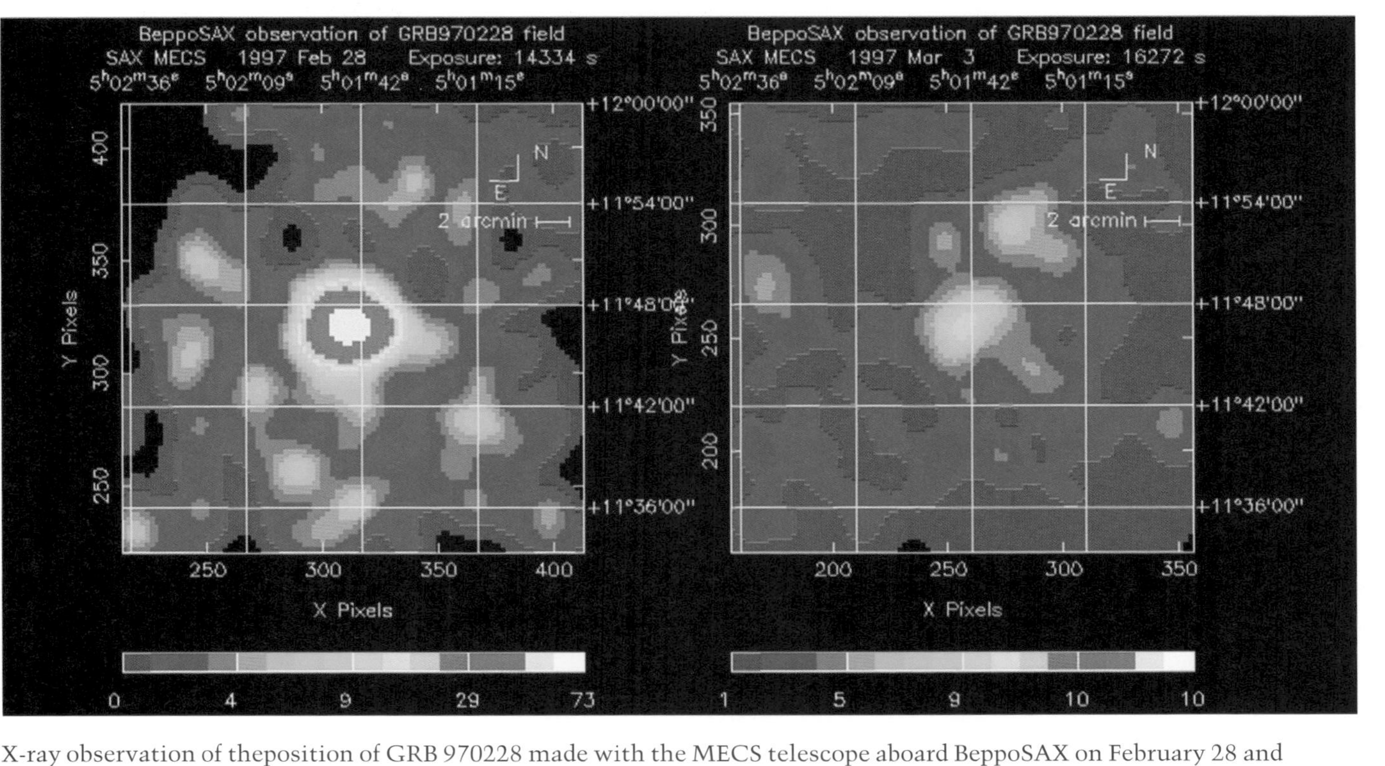

X-ray observation of theposition of GRB 970228 made with the MECS telescope aboard BeppoSAX on February 28 and March 3, 1997. The X-ray afterglow faded out almost ttally within a few days.

Close-up of the optical afterglow of GRB 970228 photographed with the Hubble Space Telescope. Besides the afterglow itself (the bright spot of light), the faint glimmer of the galaxy from which the gamma ray burst came can also be seen to the right of it.

(RIGHT) Sunrise over the dishes of the Very Large Array radio telescope in New Mexico.

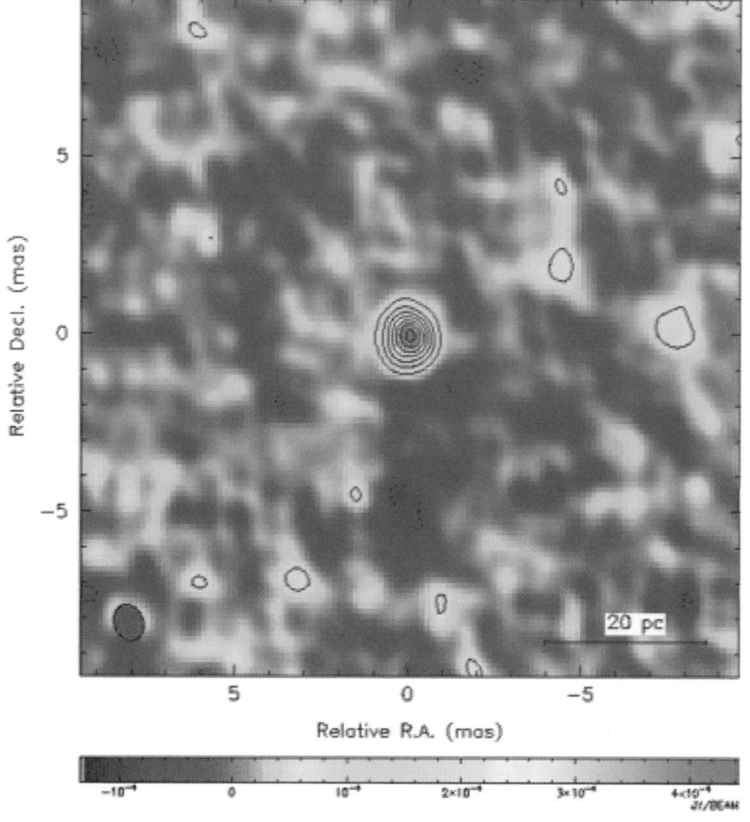

Radio emission from GRB 970508, observed with the Very Large Array.

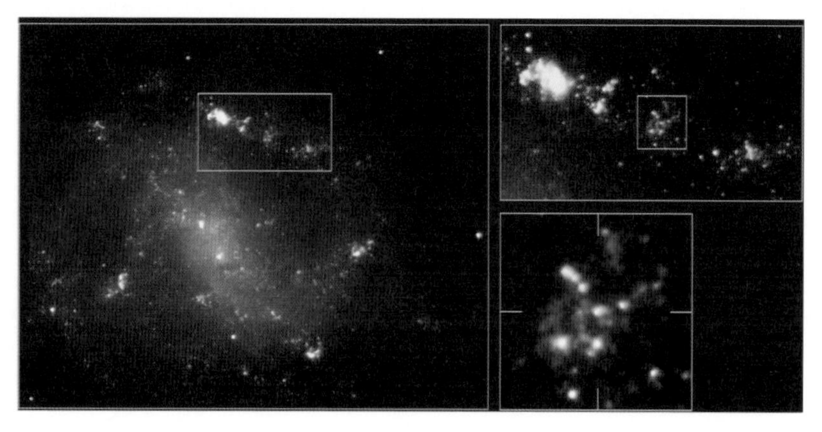

In the detailed recordings made with the Hubble Space Telescope, it can be seen that GRB 980425 took place in an active star-forming region.

Stars are born in clusters in extended clouds of gas and matter.

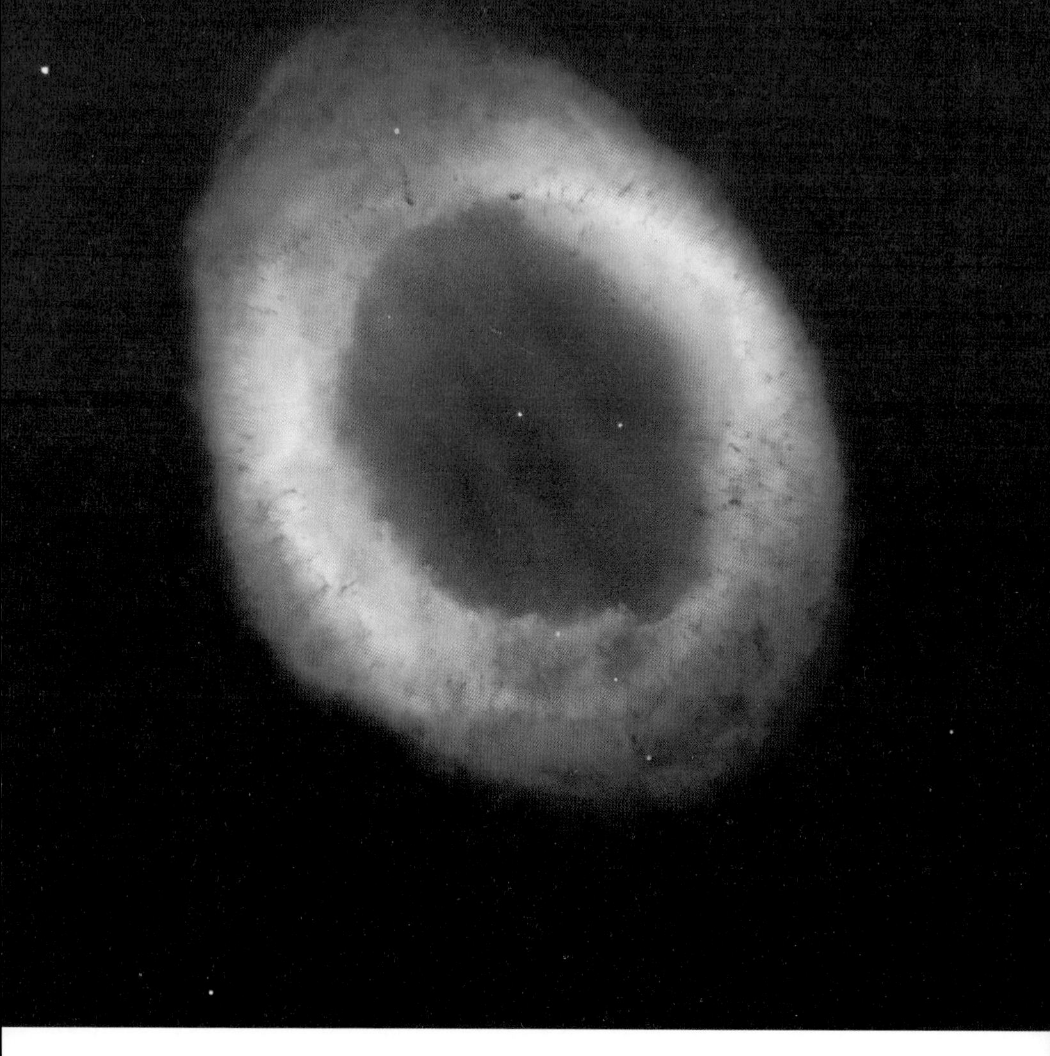

At the end of its life a star such as the sun blows into space its outer layers, which in turn form a so-called planetary nebula.

Massive stars explode at the end of their life ina powerful supernova explosion. A supernova can be just about as bright as the galaxy to which it belongs.

The Crab Nebula is the most well-known example of a supernova remnant: an irregularly formed expanding nebula that came into existance through the explosion of a massive star.

Like LOTIS, ROTSE (Robotic Optical Transient Search System) has four automatic cameras with light-sentive lenses. ROTSE is mounted at the Los Alamos National Laboratory in New Mexico.

FEBRUARY 8, 1999

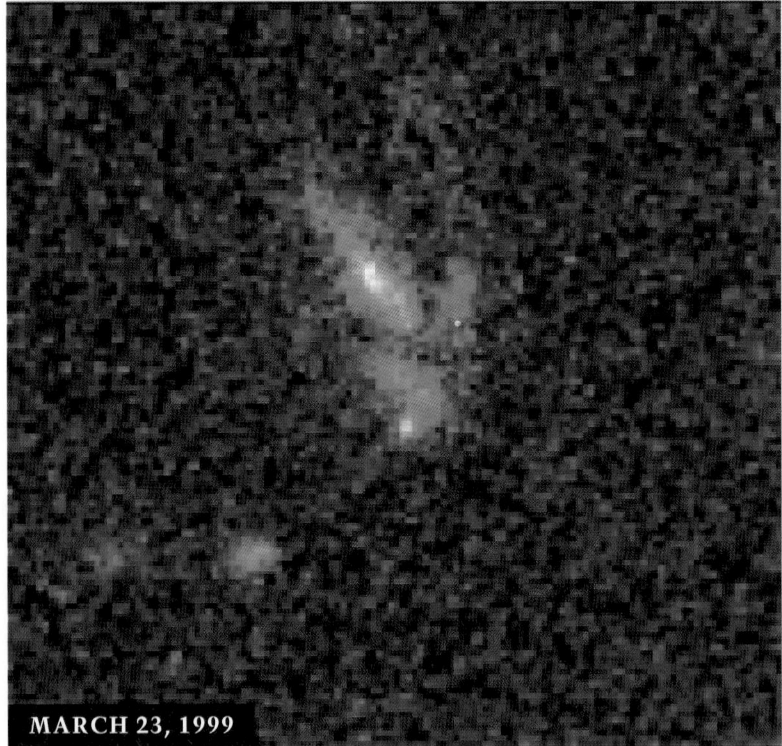
MARCH 23, 1999

Three images of the optical afterglow of GRB 990123 made with the Hubble Space Telescope on February 8, 1999, March 23, 1999 and February 7, 2000. The gamma ray burst occurred in the outer areas of an iregularly formed galaxy.

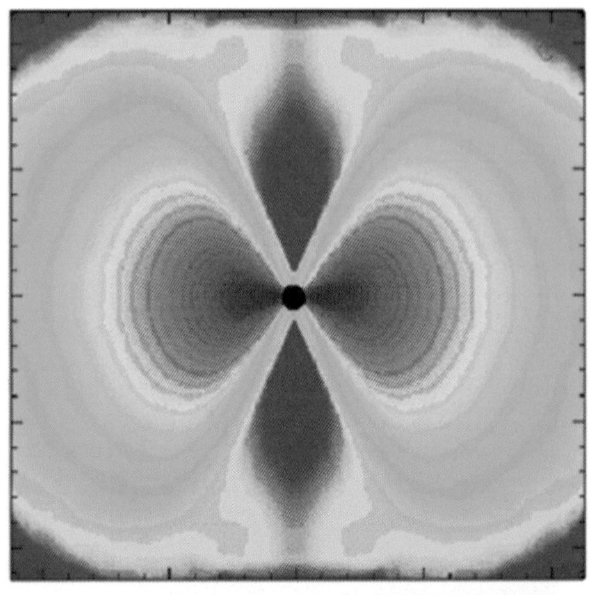

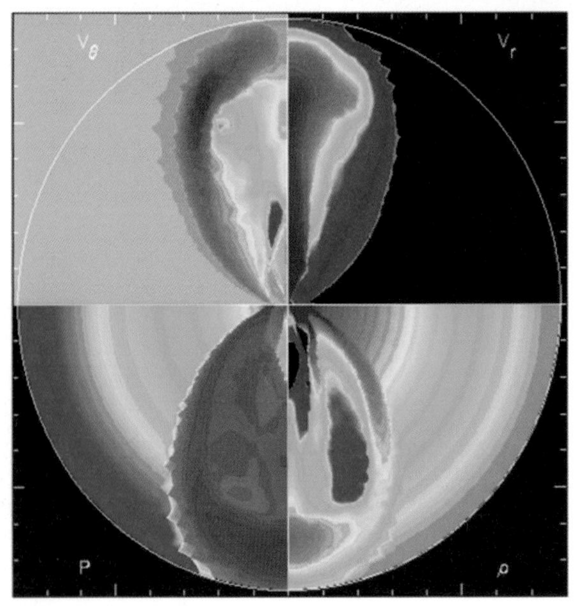

Two pictures of computer simulations by Andrew McFadyen. The top figure only sows the innermost part of the star.The differences in gas density are expressed through the use of different colors. Along the roating axis of the star the gas disappears rapidly into the central black hole. Here can be found areas with a low density (dark blue). The bottom figure depicts the situation the moment that the two oppositely directed jets reach the surface of the star (the white circle). The various colors indicate the speed (upper left and upper right), the pressure (lower left) and the gas density (lower right).

(RIGHT) Swift will research gamma ray bursts and their afterglows with a sensitive gamma detector, an X-ray telescope and a telescope for ultraviolet radiation and visible light.

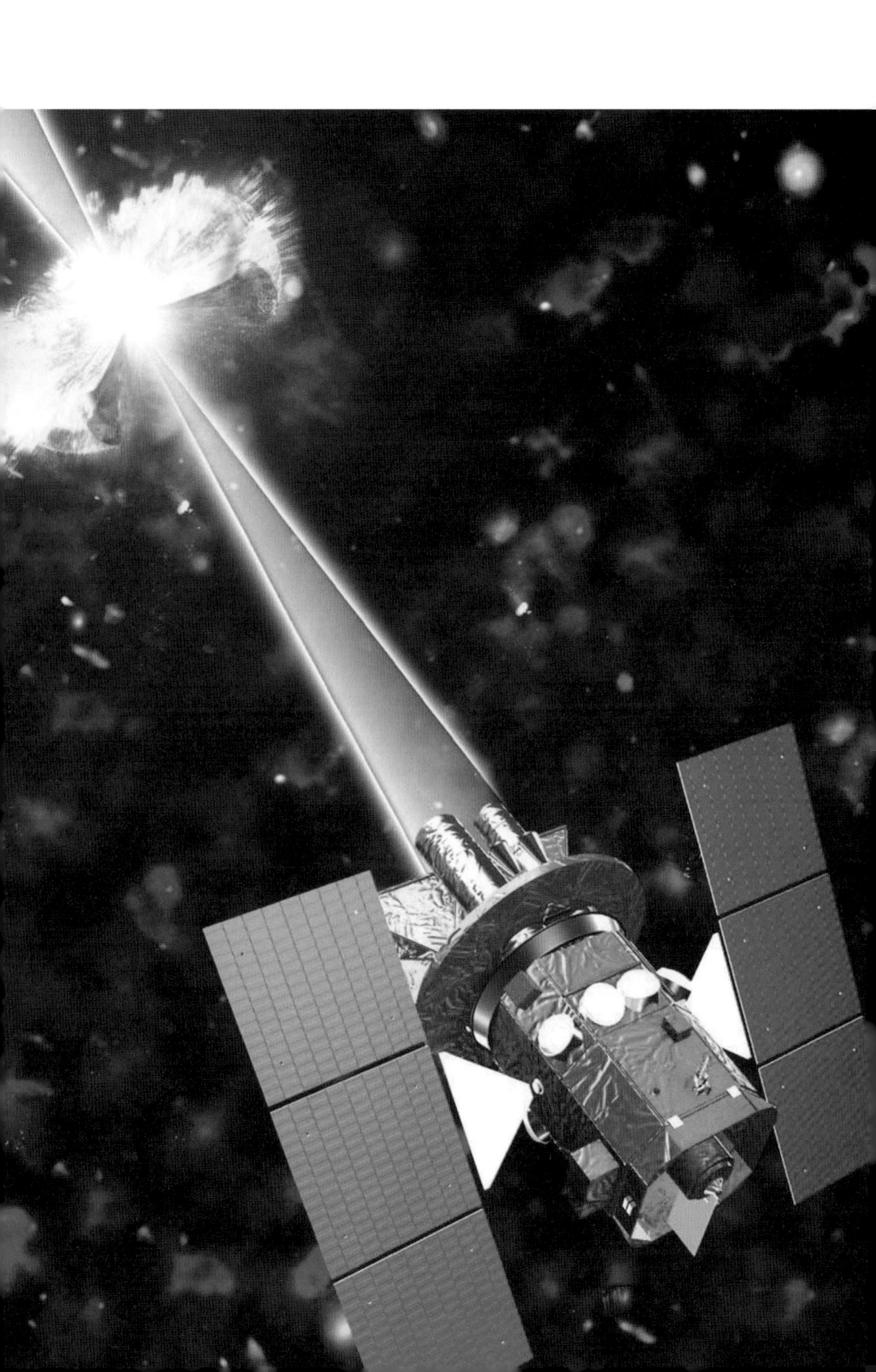

The High Energy Transient Explorer 2 (HETE 2) is the first satellite to be especially designed to research gamma ray bursts.

Schematic illustration of a gamma ray burst. If a burst should occur a short distance away in the Milky Way, and the earth (far upper right) were caught in the beam of the most powerful gamma radiation, the phenomenon could be catastrophic for life on earth.

The earth is a small vulnerable planet in a highly energetic universe.

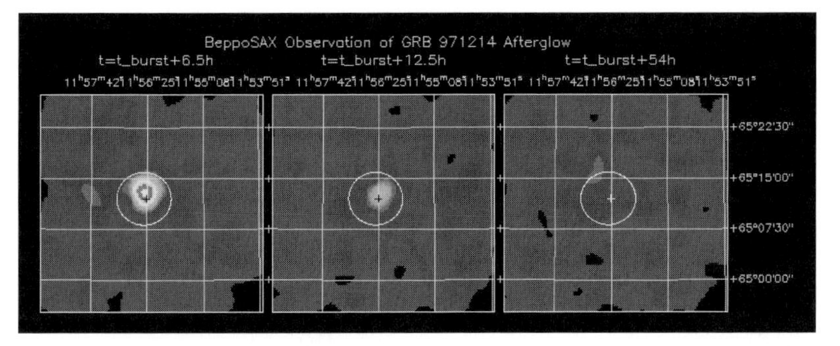

The X-ray afterglow of GRB 971214 faded away almost entirely three days after the gamma ray burst. These observations were done with the MECS telescope on board the BeppoSAX satellite.

States trained on Ursa Major: the 2.4-meter Hiltner telescope and a smaller 90-centimeter telescope on Kitt Peak in Arizona, a 1.2-meter telescope at the Whipple Observatory (also in Arizona) and the 3.5-meter telescope of the Astrophysical Research Consortium at Apache Point in the southeast of New Mexico. They all found a faint optical afterglow at the position of the X-ray source, even though there was a full moon on December the 14th.

When night finally fell in Hawaii, Kulkarni and his colleagues wanted to put one of the two Keck telescopes to work right away. Unfortunately, on that first night, they could only use a small guide camera with relatively low sensitivity. But on the next night, 37 hours after the gamma ray burst, they made the first observations with the sensitive Low Resolution Imaging Spectrograph that is mounted on the Keck II telescope. The afterglow was clearly visible on the photographs, but as time passed it got steadily weaker.

Because of the brightness of the moonlight it was not possible to get a good spectrum that first week, and a week after the burst, on December 22nd, the brightness of the afterglow had already decreased to 25th magnitude. It was not until the night of January10th, 1998, that observations could again be made. Basing his expectations on the lessening of the observed brightness, Kulkarni did not anticipate the afterglow to be any brighter than magnitude 27.5 – almost a billion times

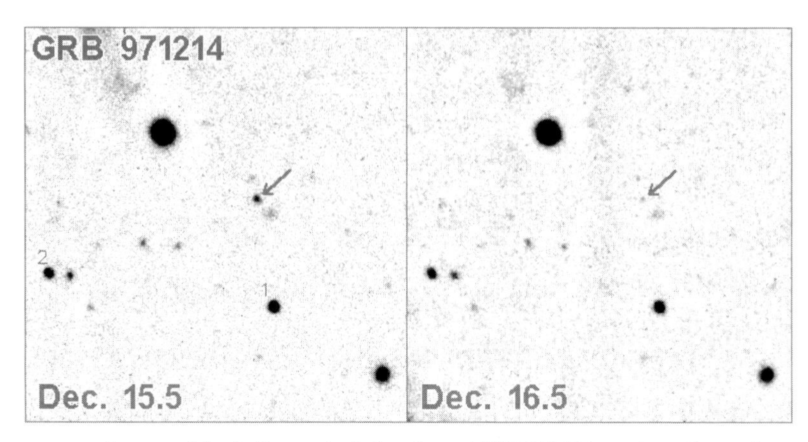

Images of the fading optical afterglow of GRB 971214, made on December 15 and 16, 1997, with the Keck telescope.

fainter than the faintest star that can be seen with the naked eye. It seemed unthinkable that a spectrum could be obtained for such a faint object.

However, on that night a small smudge of light was observed at the afterglow position; it was five times brighter than expected! This was most likely the galaxy in which the gamma ray burst had originated. It could indeed have been a coincidence that they both appeared at the same place, and a chance alignment could not be ruled out, but a quick statistical analysis showed that the probability of such an eventuality was only about 0.1%.

On February the 3rd 1998, Kulkarni was able to record the spectrum of the extremely faint galaxy. To do that, the huge Keck telescope with its enormous light-gathering power had to be aimed for hours on a barely observable spot of light. But it was worth it. In the spectrum there were several striking lines visible, among which was the famous Lyman-alpha line of neutral hydrogen gas. This line normally has a wavelength of 121.6 nanometers, in the ultraviolet part of the spectrum, but it was now found to be at a wavelength of 537 nanometers.

When a spectral line shifts from 121.6 to 537 nanometers, the wavelength shift is 415.4 nanometers, or 3.42 times the rest wavelength. To put it another way: the redshift of the galaxy was 3.42. That means that the light from the galaxy must have been traveling through the expanding universe for about 12 billion years before it reached the earth. It was sent out when the universe was almost four and a half times smaller than it is now!

A redshift of 3.42 is not a record. There are galaxies known to have a redshift of more than 5. But to observe a gamma ray burst at the colossal distance of 12 billion light years was certainly beyond expectations. This can only mean that we are witnesses to an inconceivable amount of energy. Shortly after the burst the optical afterglow was already almost fifty times as bright as the whole galaxy in which the explosion took place, and at gamma ray wavelengths there was even more energy emitted.

For many people, energy is an abstract concept. Astronomers are still using the old fashioned unit, the erg, that isn't even used in high school any more. Officially, an erg (a 10 millionth of a joule) is the amount of energy that a 2 gram mass has if it is moving at a speed of 1 centimeter per second. It is an unimaginably small amount; a Christmas tree light radiates in 1 second many, many millions of ergs. One kilowatt hour, the unit used by energy companies to calculate your bill, equals no less than 36 trillion ergs (3.6×10^{13} ergs).

Astronomers usually work with much larger amounts of energy. The sun radiates almost 4×10^{33} ergs of energy per second, as much as 4 trillion, trillion 100 watt light bulbs. During its entire lifetime of about 10 billion years the sun will produce more than 10^{51} ergs of energy, a number that defies the imagination.

It may be difficult to picture these mind-boggling amounts of energy but making such calculations is no problem, especially for astronomers. BATSE and BeppoSAX had very accurately measured the gamma ray brightness of the burst on December 14th, and now that the redshift was known Kulkarni could calculate how much energy the gamma ray burst had produced in half a minute. The result was astonishing: 3×10^{53} ergs – a few hundred times as much as the sun sends out in 10 billion years. If all that energy was sent out in visible light the burst would have been 100 million times as bright as our whole Milky Way galaxy. And in a fleeting moment of one or two seconds, the burst emitted as much energy as the rest of the entire universe.

Together with a number of colleagues, Kulkarni and Djorgovski wrote a detailed article about GRB 971214. On May 6th, 1998, one day before their story was to appear in *Nature*, NASA held a press conference in their Washington headquarters to report the amazing results. Reporters and science journalists from newspapers, magazines, radio and television were bombarded with such pronouncements as, 'The most powerful explosion since the big bang'. On that Wednesday afternoon the NASA auditorium was full to bursting.

It was indeed the most energetic explosion that had ever been observed in the universe so far. Only the big bang itself – the explosive beginning of the universe 15 billion years ago – must have been even more powerful. In an area with a diameter of one and a half kilometers, there prevailed, as a result of the gamma ray burst, the same conditions as in the newly born universe, about a thousandth of a second after the big bang, said Djorgovski at the press conference. In the next day's newspapers the gamma ray burst of December 14, 1997, got the name of *Big Bang 2*.

'The comparison of course, has its limits', says Kuklkarni. 'The big bang took place over the whole universe at the same time, while the gamma ray burst took place only at one particular position.' But Kulkarni was grateful to Djorgovski for his provocative analogy that sparked the imagination. Never before had the media and the public shown such an intense interest in gamma ray bursts as they did after the Washington press conference. 'I learned a lot that day about *public relations*', says Kulkarni. 'It is absolutely unbelievable what an enormous effect you can have just by choosing the right words.'

Big Bang 2, with its mammoth force, not only shook up the public, it even excited the gamma ray burst researchers themselves. In the middle of February 1997, most astronomers still thought that the explosions took place in the halo of our Milky Way galaxy. Speculation ran rampant: some said the explosions were huge outbursts in the immediate neighborhood of pulsars; others claimed they were the crashing of comets or asteroids into neutron stars – there were more than enough theories. But just a few months later no one questioned that gamma ray bursts were extra-galactic and were much more powerful. By the middle of February 1998, Kulkarni and his colleagues figured out that *Big Bang 2* had radiated, in less than a minute, a couple of hundred times as much energy as the sun had in 10 billion years.

For theoreticians the gigantic expenditure of energy of GRB 971214 was an unprecedented challenge. Astronomers were used to such large numbers, but even the most powerful supernova explosion in which a massive star is rent asunder at the end of its life produces no more than 10^{51} ergs of energy, which is just 0.3% of the energy of the gamma ray burst. There is no process known in nature where, in the span of less than a minute, 3×10^{53} ergs come free. Even if you could convert the 1500 trillion, trillion tons of hydrogen gas in the sun into helium all at one time, you would still come out by a factor of eight too short.

One of the popular theoretical models for extragalactic gamma ray bursts was the merging of two neutron stars, but even that model does not explain the energy produced by *Big Bang 2*. Such a rare event can produce only about as much energy as a supernova explosion. Nor does Bohdan Paczyński's 'hypernova' model of a totally annihilating super-supernova explosion of an extremely massive star offer an adequate explanation.

But theoreticians do not give up easily. Maybe gamma ray bursts actually radiate less energy – they just appear to be that bright. The idea that some gamma ray bursts look brighter than they really are had been suggested much earlier, but it seemed to be appropriate again. The magic word was 'beaming'. Maybe the energy from the explosion was radiated in a rather narrow beam and the burst looked bright only because the beam happened, by coincidence, to be pointed in our direction.

How can nature make such fools of us? It has to do with the way in which astronomers calculate the energy production of celestial bodies. Imagine that you see a star in the sky and you carefully measure how many photons per second are received on a detector of one square centimeter. You then know precisely how much energy is radiated in the direction of the detector. Naturally, just a very little bit of starlight comes on that one square centimeter because the star is also radiating in all other directions, which means that many other photons are not received on earth. To calculate the total energy production of the star you must first determine the distance. This tells you how small the detector is as seen from the star, and what minuscule percentage of the total amount of energy will finally hit that one square centimeter on the earth. A simple calculation then tells you what the total energy production of the star is per second.

But with this calculation you are assuming that the star is radiating energy equally in all directions. Strictly speaking that is not 100% definite. Maybe the 'back side' of the star is dark; maybe it radiates only in our direction. That may be unthinkable for a star, but for a gamma ray burst we really don't know. After all, it must be borne in mind that almost nothing is known about the nature of gamma ray bursts.

Imagine that you are standing in a field on a dark night. One kilometer away there is an apartment building and behind one of the windows on the top floor you see a light. It comes from a bare bulb on the ceiling in the room of your colleague. From the apparent brightness of the light and the known distance to the building, you can figure out how much energy the light is radiating per second. That also gives you the power of the light bulb, for example 100 watts.

Now, when a light of equal brightness shines from the dark window next door it seems obvious that a 100 watt bulb is turned on in that room too. But things are not always what they seem. The following day your colleague tells you that what you saw in the other room was actually a flashlight with a measly 6 watt bulb. But because the reflector and the lens of the flashlight concentrate the light totally in one direction, and in addition, it had been aimed right at you, the brightness was seen as being equal.

Who knows whether or not gamma ray bursts such as GRB 971214 come from cosmic flashlights and actually produce much less energy than was first thought? When the energy is sent out as a beam and the

Hubble image of a powerful jet that is ejected from the nucleus of the galaxy M87. Similar relativistic jets also play a role in gamma ray bursts.

beam happens to be pointing towards the earth, then the energy production of gamma ray bursts may be less prohibitive. It is as if you hear someone yelling a few kilometers away even when you know that the human vocal chords cannot produce that loud a sound. But when you see that a megaphone is being used then you have the answer to the puzzle. The beaming of the voice through the megaphone carries the sound further and makes it come across much louder even though the actual volume of the voice is less.

Astronomers are aware of many examples of objects that show this flashlight or megaphone effect. In most cases there are two beams in opposite directions more-or-less along the rotational axis of the object or along its magnetic axis. Rapidly rotating pulsars sweep lighthouse-like beams of radiation through the universe. Young protostars blow material out in two directions. Dying stars often shroud themselves in two-lobed expanding clouds of ejected stellar gas. Active galaxies and quasars produce strong jets of fast moving electrically charged particles. It would actually be very surprising if gamma ray bursts did not show beaming.

If gamma ray bursts do indeed radiate their energy in two opposite beams, each spreading with an angle of some 15 degrees, then the 'energy crisis' from explosions such as *Big Bang 2* no longer presents a problem. Seen from the center of the explosion the two beams together cover only 1% of the sky. To put it differently: if observers on earth should by coincidence find themselves looking into one of the two beams, they will overestimate the total energy expenditure of the explosion by a factor of 100. In that case the December 14, 1997, energy production was not 3×10^{53} ergs but only 3×10^{51} ergs, which can be compared to the energy of a supernova explosion, and which is in agreement with the theory of the merging of neutron stars.

It is an intriguing idea but there is also a downside. If we on earth only see those gamma ray bursts that happen to have the right orientation, then there must be many more gamma ray bursts in the universe than have been observed up until now. For each burst from which one of the beams happens to be aimed at the earth, there are a hundred others we cannot see. The BATSE detector on board the Compton Gamma Ray Observatory detected about one burst a day, but when beaming plays a role then the frequency lies more in the area of one every 15 minutes. The mysterious explosions may be a hundred times less luminous but they would also be a hundred times more numerous.

That also means the total amount of energy that is produced over time by all the gamma ray bursts together is exactly the same. If the radiation is beamed, the amount of energy per burst is 100 times smaller, but because the frequency of the bursts is 100 times greater, then the total amount of energy is not affected. If, on a particular day, there is either one very powerful gamma ray burst or one hundred 'tiny' ones, the fact remains that on the average the same amount of energy is produced per day by a handful of short gamma ray bursts as there is by 10 billion stars such as the sun.

However you look at it, *Big Bang 2* got the theoreticians busy refining their models. Only if the gamma rays from the explosion were in some way beamed, is it possible to explain the observed brightness of the far-removed bursts. Beaming became the newest 'buzz' word of the gamma ray burst community, and very quickly there appeared the first detailed model calculations and computer simulations from which it seemed that the forming of jets could hardly be avoided.

But the powerful burst on the 14th of December 1997 gave observers a jolt. GRB 971214 was only the second burst for which the redshift and the distance were determined and immediately astronomers had to start dealing with an object more than 10 billion light years away. It seemed likely that many more gamma ray bursts at enormous distances would be discovered with redshifts of perhaps even 4, 5, or 6, which would correlate with still greater distances. Indeed, on 31 January, 2000, astronomers detected a gamma ray burst with a redshift of 4.5. At such large distances, gamma ray bursts could be used to study the early youth of the universe.

When you look at an object a great distance away in the universe, you are also looking back in time. The light of *Big Bang 2* took some 12 billion years to cover the distance to the earth. And since it was generated when the universe was only 2 billion years old, a study of the gamma ray burst can also reveal more about the early universe.

'After GRB 971214, many astronomers began to think about using gamma ray bursts as a cosmological tool', says Kulkarni. 'In the future such an application could well become very important.' It wouldn't be the first time that puzzling objects in the universe delivered a wealth of new information for other areas of astronomy. Pulsars are a good example. The first pulsar was discovered in 1967 and now more than a thousand are known. They are super-fast spinning neutron stars that send out, with the precision of an atomic clock, short pulses of radio waves or X-rays. Today, the pulsar mechanism is quite well understood but long before that happened pulsars were 'used' to study interstellar material, the physical properties of their companion stars, and the theory of gravity.

For quasars, which are probably the active nuclei of very distant galaxies, the story is more-or-less the same: how the energy of a quasar is produced is still not well known, but by studying quasars astronomers have come to know a great deal about the origin of galaxies and clusters in the early time of the universe and about the distribution of thin hydrogen gas clouds in intergalactic space.

According to Kulkarni, something similar could happen with gamma ray bursts, and to a certain degree, it is already happening. The fact that distant gamma ray bursts appear to be associated with galaxies offers valuable information about the evolution of the universe. A burst like *Big Bang 2* with a redshift of 3.42 happened 12 billion years ago when the universe was just 2 billion years old. That a faint nebula was found at the position of the burst makes it clear that 2 billion years after the big bang, galaxies already existed. In the future, when gamma ray bursts are detected with higher redshifts, the history of the origin of galaxies will be completely mapped. Without the luminous beacons from the gamma ray bursts, such extremely faint galaxies would never be discovered.

Moreover, the spectrum of the optical afterglow of a gamma ray burst can be searched for in the absorption lines of the neutral hydrogen gas that is found in the space between the burst and the earth. The distances of these hydrogen clouds may then also be determined and in this way astronomers get a picture of the distribution of material in the space between the galaxies.

The year 1997 was a revolutionary one for gamma ray burst research. For the first time the optical afterglow was discovered; for the first time the redshift and the distance could be fixed; for the first time radio astronomers confirmed the popular relativistic fireball model and for the first time theoreticians became aware of the need to develop models for asymmetric explosions. In three cases an optical afterglow was discovered and all three seemed to coincide with a dim, distant galaxy. From follow-up observations on these very faint galaxies, using the Hubble Space Telescope as well as others, it became clear that a great number of new stars were formed in these galaxies, suggesting a relation between gamma ray bursts and star-forming regions.

Could that mean that gamma ray bursts are indeed related to extremely powerful supernova explosions? In an area where many new stars are born, there are always several massive, hot ones that never survive more than 10 million years. While the formation of stars in their immediate neighborhood goes on unremittingly, these short-lived massive stars end their brief existence in a gigantic supernova explosion. Every once in a while a hypernova may even take place in which an extremely massive star collapses completely into a black hole and throws high-energy gamma rays out into space in two directions.

'All the issues that Bohdan Paczyński and I raised a year ago have now been resolved', wrote theoretician, Ralph Wijers in the spring of 1998 in *Nature*. 'It might seem optimistic to think that all the [new] issues raised here will once again be solved within a year but the pace of discovery is still very rapid.' Wijers's optimism turned out to be completely justified. While his article was still lying on the desk of the *Nature* editors, his colleagues in Amsterdam got the first convincing indications that there is indeed a connection between gamma ray bursts and supernovae – not at a distance of 12 billion light years away, but to everyone's amazement, in a galaxy right around the corner.

10 Curious connections

For you see, so many out-of-the-way things had happened lately that Alice had begun to think that very few things indeed were really impossible.

The name of Anna Belomo never appeared in the pages of *Nature*, but at many gamma ray burst conferences the Anna correlation (named for her) was often a subject for discussion. One time, when she was particularly irritated, she remarked that those damn gamma ray bursts always happen on the weekends or holidays. How many evenings and romantic nights were messed up because the pager of her boyfriend, Titus Galama, went off, urgently calling him to get over to the astronomy institute immediately – and there he stayed, out of sight, for the next 2 or 3 days?

Galama, the scientist through and through, decided to make a graph, placing the days on a horizontal line and the number of gamma ray bursts in the vertical columns to see if there was any validity to Anna's complaint. And sure enough, the graph showed remarkable peaks on Friday, Saturday and Sunday. Every time the Anna correlation, with its unexplainable relationship between the days of the week and the mad dash to record afterglows, was put on an overhead projector and shown at meetings, it invariably broke the audience up with laughter.

That her name would forever be associated with one of the greatest enigmas in astrophysics is something Anna could not have imagined 10 years ago. In her birthplace of Barcelona she might have romantically looked at the moon and the stars with a boyfriend, but those heavenly bodies began to twinkle in a different way when she happened to meet three Amsterdam astronomy students who were on vacation in Spain. Titus with his long lean body and luxuriant head of dark hair was far and away the best looking of the trio. A year later Anna joined him in Amsterdam in his tiny flat on the edge of an old, traditional working-class neighborhood called 'de Pijp'. A couple of years after that, on January 11,1997, BeppoSAX came up with the first accurate position of a gamma ray burst and Titus' chase after afterglows was rewarded. Of course it was on a Saturday.

Unexpected happenings always come at inconvenient times and it is no different for gamma ray bursts. Cosmic explosions have no respect for clocks or calendars. The Gamma Ray Burst Monitor and the Wide Field Cameras of BeppoSax are working 24 hours a day and when they happen to succeed in fixing the position of a gamma ray burst they don't synchronize with office hours or the sleeping patterns of earthly astronomers. Enrico Costa, who is usually the first to drag his colleagues in America and Europe out of bed, knows all about it. 'All the women must hate me', he says, grinning.

Costa's colleague Marco Feroci had just come back from a conference in the United States on May 8, 1997, when a gamma ray burst flashed. After a much too short and sleepless night on board a plane and a bad case of jet lag, Feroci ended up working for 36 hours. For Paul Groot in Amsterdam this was one of the reasons that he later decided to work on something other than gamma ray bursts. 'I really need to get my sleep.'

Alberto Castro-Tirado from the LAEFF institute in Madrid was called by Costa late at night on the 28th of August, 1997, and he immediately raced up to the Calar Alto observatory in the Sierra Nevada a few hundred kilometers to the southeast. That earned him a couple of speeding tickets but even worse, when he got to the observatory he found that he had forgotten his key to the control room. It took a few acrobatic tricks and contortions but he finally reached the telescope and got it working only to discover that at the place of the gamma ray burst there was no optical afterglow to be seen.

Luigi Piro, project scientist of BeppoSAX, was visiting his cousin on July 3,1998, to watch the world football championship game between Italy and France. His mobile phone was broken and when, in the middle of the game, his cousin's telephone rang, he knew for certain that there must be a new gamma ray burst – nobody else would dare call at that time. He never saw the end of the game. Italy lost 3-4.

Joshua Bloom, one of Shri Kulkarni's graduate students in Pasadena, had come home from a party on the night of January 22–23, 1999, and had just dropped into bed when the phone rang to tell him about GRB 990123. Half drunk, with the telephone wedged between his shoulder and his ear, he pulled on his clothes while he staggered to his car. It wasn't until sometime later that he discovered he had put his pants on backwards.

Always inconvenient. That was also true for the gamma ray burst on April 25th, 1998. There was nothing unusual about it. It was of average brightness and duration – and of course it happened on Saturday night (21.49 Universal Time to be precise), completely confirming and conforming to the Anna correlation. But there rode Titus Galama on his bike late at night to the Anton Pannekoek astronomy institute at the University of Amsterdam where his colleague Paul Vreeswijk, successor to Paul Groot, was already waiting. Another weekend shot.

GRB 980425 was located in the southern sky in the small and unremarkable constellation called Telescopium, named by the eighteenth-century French astronomer Nicolas de Lacaille for the most important instrument in astronomy. Neither the United States nor Europe was able to observe that position in the sky and even the observatory on La Palma was not far enough south. Luckily, in 1997, van Paradijs's team had also arranged override proposals for telescopes in the southern hemisphere such as the Anglo-Australian Observatory on Mt. Stromlo in New South Wales and the European Southern Observatory (ESO) in Chile.

From their small office on the third floor of the astronomy institute in the east end of Amsterdam, Galama and Vreeswijk began to organize as many observations as possible in the hope of tracking down an optical afterglow. So far, astronomers had been able to quickly determine the sky position of a gamma ray burst in twelve cases, but only in five of them had they been successful in finding something with an optical telescope. All of these optical afterglows were exceptionally faint but speed was of the essence; the sooner an afterglow could be observed, the better.

Unfortunately the big 3.9-meter Anglo-Australian Telescope was not available on the night of April 26–27. It was being used for spectrographic observations and it was impossible to quickly replace the spectroscope with a camera. Luckily, a smaller 1.25-meter telescope on Mt. Stromlo, which is usually used to search for dark matter in the Milky Way galaxy, was available. The first images of the burst position were made on Sunday afternoon as soon as night had fallen again in Australia and the constellation Telescopium appeared above the horizon.

Not much could be done with these first images. The error box of the Wide Field Camera of BeppoSAX had a diameter of about 15 arcminutes – half as big as the full moon – and in this field there were countless faint stars. But in a day or two, after images would have been made with which to compare, it would be possible to see if, out of all those many little stars, one had in the meantime become fainter.

Two days later observations with the 3.5-meter New Technology Telescope (NTT) at the ESO observatory on top of La Silla in the Chilean Atacama desert could also finally be made. The NTT, which went into service in 1990, is one of the most modern medium-size telescopes in the world, equipped with revolutionary technical extras and advanced observational instruments. In fact the NTT, at that time, was built as a small-scale prototype of the European Very Large Telescope consisting of four 8.2 meter telescopes. With its sensitive detectors it had to be possible to get a fix on a very faint afterglow in a relatively short time.

While all this was going on, the Italian BeppoSAX team was also on its toes. Within the error box of the Wide Field Camera, two X-ray sources were found with the help of the LECS and MECS telescopes, one of which would probably be the X-ray afterglow of the gamma ray burst.

On Wednesday the 29th of April, when Galama and Vreeswijk received the NTT images to study, they decided for safety's sake, to compare them with the electronic archives images from the Digitized Sky Survey (DSS), which only shows relatively bright stars. This had been done before, to no avail, but they thought it wouldn't hurt. It immediately became clear that something very unusual was going on. In the original error box of the Wide Field Cameras they saw that there was a rather bright star visible on the NTT images but it was missing from the old DSS images. The star was in the outer edge of a spiral galaxy and was 16th magnitude.

Could this be the optical counterpart of GRB 980425? That seemed highly improbable; it was almost 1000 times brighter than the ordinary afterglow. Of course it could mean that this gamma ray burst had been much closer, but then the gamma ray luminosity would also have had to be 1000 times brighter. Could another possibility be that it was an unrelated supernova in one of the spiral arms of the galaxy? Even though it looked like that, such a coincidence of circumstance was more than a little suspect. The chance that two different explosions had taken place at approximately the same moment and at approximately the same place in the sky is negligibly small.

Galama and Vreeswijk decided to make their remarkable finding known to the world as quickly as possible even though they were not quite certain how the whole thing fitted together. Early on Wednesday evening, after discussing it in detail with colleagues in Amsterdam, Rome and Huntsville, they sent a brief, factual e-mail to the GCN (Gamma-ray burst Coordinates Network), an electronic service that had recently been set up to inform interested researchers about observations of gamma ray bursts as quickly as possible.

The GCN is the brain child of Scott Barthelmy of NASA's Goddard Space Flight Center. In the summer of 1993, Barthelmy had set up a computer system for the dissemination of burst positions from BATSE. From the telemetric data of the Compton Gamma Ray Observatory a fully automatic selection was made of the suspect signals, and by comparing the signal strength on the eight different BATSE detectors, it was possible (also automatically) to derive a rough estimate of the sky position. A couple of seconds after receiving the observations from the Compton Gamma Ray Observatory, the information was dispersed by e-mail over the entire world.

The arrow points to supernova 1998bw in the galaxy ESO184-G82. The supernova explosion occurred almost at the same time and at the same position in the sky as the gamma ray burst 980425.

The electronic GCN circulars, which are actually an outgrowth of this BACODINE project (BAtse COordinates DIstribution NEtwork), first appeared in August 1997. They were haphazardly numbered in the beginning, but later on Barthelmy began to list the circulars systematically. The discovery of the mysterious explosion by Galama and Vreeswijk was described in circular 60 and by the end of 2001 the number of GCN circulars exceeded 1100. For gamma ray burst researchers they serve as a superb communication platform.

Very quickly it was apparent that the new star was indeed a supernova. The IAU (International Astronomical Union) designated it SN 1998bw. During the next several weeks the brightness increased until it reached magnitude 13.8 on May 13th. Thereafter, little by little, it became more and more faint, behaving just as astronomers would expect a supernova to behave. The galaxy in which the supernova appeared is known to astronomers as ESO 184-G82; its distance was calculated to be 140 million light years – not exactly around the corner but much nearer than the many billions of light years that gamma ray burst researchers had grown accustomed to.

Meanwhile, the coordinates from the two X-ray sources from BeppoSAX were also published, and neither of the sources turned out to coincide with the supernova. Apparently, it really was a chance coincidence. Or was it? Galama and Vreeswijk and some of their colleagues studied the statistics intensely and came to the conclusion that the chance of encountering a relatively luminous supernova in one of the thirteen BeppoSAX error boxes is only one hundredth of a percent. And furthermore, this was certainly not a normal supernova.

In the first week of May the supernova was studied in detail by Australian radio astronomers. They discovered that SN 1998bw produced an extaordinary amount of radio waves. In addition, spectroscopic observations carried out by astronomers from the ESO observatory indicated that this particular supernova was a type that is seldom seen. Finally, a very careful analysis of the course of its luminosity showed that the actual explosion took place somewhere between the 21st and 27th of April, consistent with the time of the gamma ray burst.

And what about the two X-ray sources from BeppoSAX? As it turned out, they were not a stumbling block. A new analysis of the LECS and MECS observations indicated that one of the two sources was actually not there at all, and after the observations had been corrected for a slight positioning problem of the satellite, it became apparent that the second X-ray source coincided precisely with the supernova. In the middle of May most astronomers had little doubt that supernova SN 1998bw and gamma ray burst GRB 980425 were one and the same object. The only alternative, which seemed very unlikely, was that right next to that peculiar supernova, and almost exactly at the same time, a gamma ray burst, unobservable at any other wavelength, had also occurred.

If the remarkable supernova that Galama and Vreeswijk discovered had produced a gamma ray burst, then the gamma ray burst itself was also unusual. While previous bursts took place many billions of light years away, GRB 980425 was at a distance of only 140 million light years. The fact too that it had not been very powerful despite the small distance, could only mean one thing: this gamma ray burst must have produced a few hundred thousand times less energy than normal! 'If one accepts the possibility that GRB 980425 and SN 1998bw are associated, one must conclude that GRB 980425 is a rare type of gamma ray burst and SN 1998bw is a rare type of supernova', said Galama and his co-authors in an article in the October 15th, 1998, issue of *Nature*.

In the same issue of *Nature* there was a detailed article about the unusual radio properties of the supernova and a theoretical discourse on stellar explosions, written by an international group of astronomers under the leadership of Kohichi Iwamoto from the University of Tokyo. According to this group, this was really a 'hypernova' with an explosion energy of perhaps ten times greater than a normal supernova. Judging from the spectrum it appeared that the exploded star was primarily made up of carbon and oxygen. Apparently, it had blown off its mantle of hydrogen and helium in an earlier stage of its evolution.

By comparing the spectrum and the light curve of SN 1998bw with various other model calculations, it was possible to derive the properties of the exploded star. Without question, the best match was obtained when you assumed that the star was at least fourteen times as massive as the sun and the explosion energy was 3×10^{52} ergs. In the beginning, the star must have been even more massive before it unloaded its hydrogen and helium, possibly 40 times as massive as the sun.

The calculations of Iwamoto's team also indicated that the explosion must have left behind a collapsed stellar core almost three times as massive as the sun. Normally, after a supernova explosion, a neutron star remains (see Chapter 11), but that could not be so in this case. Since neutron stars weighing as much as three solar masses are unstable, there seemed to be only one possibility left: the hypernova must have created a black hole. Apparently, Titus Galama and Paul Vreeswijk had been privileged witnesses to the birth of a black hole – an object with such a powerful gravitational field that even light can not escape from it.

The destruction in that distant galaxy ESO184-G82 must have been total. In an infinitesimal fraction of a second 6000 trillion, trillion tons of stellar gas were sucked in through the one-way door of a black hole while the dying star blew even more gas out into the universe almost at the speed of light – an inferno of run-away nuclear reactions, shock waves and scorching radiation. Should such a catastrophe occur in the immediate neighborhood of our sun, there would be little left of life on earth.

The gamma ray burst of April 25th, 1998, was the first for which a positive relationship was established between the death of massive stars and the birth of black holes. That gamma ray bursts were the birth cries of black holes had been postulated years before by Bohdan Paczyński from Princeton and Stan Woosley from the University of California at Santa Cruz (see Chapter 14). According to them, a supermassive star collapses completely in on itself at the end of its life while 'ordinary' massive stars, for the most part, explode.

The discovery of the 'supernova gamma ray burst' beautifully supported Paczyński's and Woosley's theories, but it was not at all certain that this model could explain all gamma ray bursts. The fact that the gamma ray luminosity of GRB 980425 was so incredibly low compared with other gamma ray bursts made astronomers suspect that in this case they were dealing with a completely new mechanism. According to Galama and Vreeswijk in their *Nature* article, gamma ray bursts are apparently produced in different ways, and on the basis of the observed properties it was not possible to distinguish one from the other. As a matter of fact, when GRB 980425 was first observed, it looked quite ordinary.

At the 192nd meeting of the American Astronomical Society (AAS), which was held in early June in San Diego, California, the exciting discovery by Galama and Vreeswijk was the hottest topic of the day. At a crowded press conference Woosley presented his own 'collapsar' theory (his name for a collapsing star that completely implodes), saying that he was 90% convinced that GRB 980425 was the long awaited missing link.

Galama sat in the back of the room while Woosley enthusiastically talked about his theory. Of course, Galama would rather have been the one talking about his own discovery, but because he had submitted an article to *Nature* he was not yet supposed to 'go public'. But when Chryssa Kouveliotou told the journalists that the Amsterdam astronomer who had made the amazing discovery was right there at the meeting, they all crowded around him and wanted to know every detail.

Suddenly, Titus Galama was the center of attention. His name appeared in all of the most important newspapers and magazines and he was invited for interviews on radio and television. That this was happening to him is something Galama wouldn't have dreamed of a couple of years earlier. Ever since he was a small child playing in the sand box, he had been an observer; he would watch everything that was happening around him before he would become involved. His father was a virologist, and in high school Titus thought that he would go on to study biochemistry. But then, one day, his mother brought home the book, *A Brief History of Time*, by Stephen Hawking. Titus devoured it from front to back. He was completely caught up in the fascinating world of modern physics and decided to study physics in his home town of Amsterdam.

That he was lured into astronomy after his first year at the university had to do with the inspiring introductory course in astronomy given by Ed van den Heuvel, director of the astronomy institute. Supernovae and the big bang seemed even more interesting than particle physics. It wasn't long before he joined the research group of Jan van Paradijs and landed up in the energetic world of gamma ray bursts. And now, here he was in San Diego, a young Dutch PhD student with two *Nature* articles under his belt of which he was the first author, standing in front of a crowd of American science journalists, explaining to them how it was possible that such an important breakthrough could be made by someone from an out-of-the-way place like Holland.

How important the breakthrough really was, remained unclear. Was this really a completely different type of gamma ray burst? Could it be possible that 'normal' bursts are also caused by exceptional supernova explosions? Is each gamma ray burst the birth cry of a black hole? How often do such 'hypernova' explosions take place? Questions, questions and more questions, but no one knew the answers.

According to Woosley it is highly unlikely that the GRB 980425 was an 'ordinary' gamma ray burst that happened to be seen from the side. If the high-energy radiation from a gamma ray burst is concentrated in two oppositely directed jets and the earth is not in line with either of the beams, then one would indeed expect to observe a much weaker signal. However, you would not see a short burst but rather, a much longer-lasting explosion. 'In this case the total amount of energy that was radiated at gamma ray wavelengths was considerably smaller than the energy in a visible supernova explosion,' says Woosley. 'It was an intrinsically weak burst that only happened to be observable because it was not far away. In actuality, these kinds of weak bursts might well be more numerous than the powerful bursts that are observed at distances of billions of light years away.'

It wasn't too long before astronomers began to track down additional possible associations between gamma ray bursts and supernovae. And this time it did not have to do with weak bursts in the immediate surroundings of the MilkyWay galaxy, but with bursts at distances of billions of light years. If you knew what to look for, you would find supernovae hiding in the light curves of some 'normal' gamma ray bursts too.

Joshua Bloom, one of Shri Kulkarni's graduate students, was the first one to make that suggestion. Joshua had been interested in science and technology since he was a little boy. In the basement of his parent's house, on the east coast of the United States, he took radios and television sets apart just to see how they worked. He probably got his interest in the exact sciences from his grandfather, who was a psychologist and an 'ardent fan of Albert Einstein', according to Bloom. Josh started college as a physics major at Harvard but quickly changed to astronomy, just as Titus Galama had done in Amsterdam.

Purely by coincidence, he found himself in gamma ray burst research. 'In the summer of 1993 I had a job as cook on a whale watch boat for tourists', he relates. 'It was nice but not terribly inspiring. The next year I definitely wanted to do something else. Through a friend of mine I was offered a job for 3 months at the Los Alamos National Laboratory in New Mexico where I came in contact with gamma ray bursts through Ed Fenimore.' Bloom found Fenimore so inspiring, he wrote to his parents that he had the best summer job in the whole world and he was convinced that Fenimore would win a Nobel prize. Later, he worked for a year with Martin Rees in Cambridge and then in the summer of 1997 he left for Caltech to be part of Kulkarni's team. 'To tell the truth, when I first met Shri I thought he was a little weird, but his enthusiasm was contagious.'

Bloom admits that his discovery was a chance happening. 'But at Caltech we are good at planned serendipity', he says with a laugh. 'We observe interesting objects for a long time because you never know what else you might find.' The faint afterglow of GRB 980326, which had first been seen by van Paradijs's group in Amsterdam, was just that sort of an interesting object, with a brightness that decreased amazingly fast. For a few weeks the afterglow was observed with the big telescopes in Australia and Chile and with the Keck telescope in Hawaii, until around the middle of April, when nothing was left but a tiny point of light that no longer seemed to get dimmer. That had to be the far distant galaxy in which the burst took place. Obviously, the afterglow had become so faint that the enduring glimmer of the galaxy became visible.

Even the colossal Keck telescope was not sensitive enough to record the spectrum of that extremely faint galaxy. As a result, the redshift too could not be determined, so the exact distance was not known. But the Caltech group wanted to write an article about this 'host galaxy', and in December 1998 they planned to make new observations with the Keck

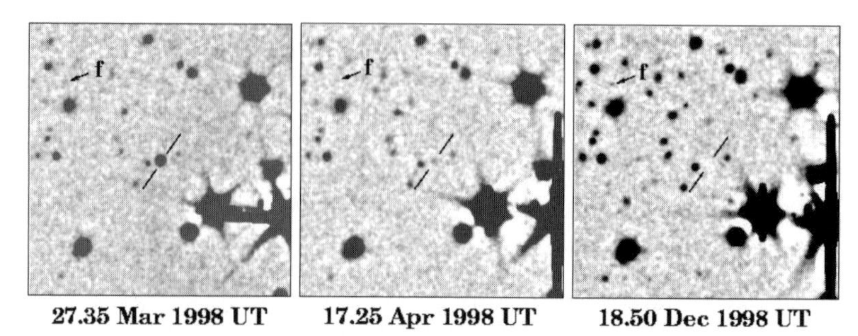

Three images of the optical afterglow of GRB 980326, made on March 27, April 17, and December 18, 1998. The weak object in the second photograph was originally believed to be the galaxy from which the burst came, but in the third picture it is nowhere to be found. It is probable that in April, light from a supernova was detected.

telescope. 'Together with Shri I went to Hawaii', says Bloom, 'and we analyzed our observations right there. To our great surprise, the galaxy had disappeared!' Astronomers are used to having the unexpected happen, but for a galaxy to just pick up and vanish went a little too far. The only explanation that made sense was that the faint point of light that had been observed in April was not the host galaxy at all but was still the afterglow of the gamma ray burst.

The big question then was, why did the afterglow seen in April not get any dimmer? 'We began by re-examining the observations made in March and April', says Bloom. 'It was only later, after hearing a talk by Stan Woosley, that I came up with the idea that maybe a supernova was playing a part here.' The optical afterglows of gamma ray bursts reach their maximum luminosity at, or just a little after, the explosion. In contrast, the visible light from a supernova does not reach its maximum brightness until a few weeks after the explosion. Now just imagine that GRB 980326 also coincided simultaneously with a supernova. A few weeks after the gamma ray burst, while the optical afterglow was rapidly becoming fainter, the light from the supernova would be reaching its maximum luminosity. The decrease in luminosity of the afterglow could possibly be compensated by the increase in the luminosity of the supernova, with the result that the faint little point of light kept its brightness for some time longer. After a while, when the light from the supernova was also extinguished, the diminutive little star definitely disappeared from sight.

Bloom and his colleagues wrote up their observations and theories in an article that was submitted to *Nature* on March 23rd, 1999, and published on the 30th of September. But at the end of May they also put their paper on the Internet so that other researchers would have the opportunity to take another look at their own observations. One month later, Daniel Reichart from the University of Chicago, a student of Don Lamb, came out with a new analysis of the light curve of GRB 970228, the first gamma ray burst for which an optical afterglow had been found. According to Reichart, here, too, a supernova might have caused a remarkable 'delay' in the decrease of brightness of the afterglow.

Independent of Reichart, Galama had also gone back to study the observations of GRB 970228 and came to the same conclusion. Back in 1997, in his *Nature* article about the first afterglow, he had already pointed out the deviant behavior of the light curve. Normally, the gradual loss of luminosity follows a so-called power law in which the decrease is fast in the beginning and then slows down, comparable to the way the sound of a gong dies away. Such a power law is characterized by only one coefficient that describes the speed of the diminution. But the afterglow of GRB 970228 did not play strictly by the rules. In the first week of March it remained brighter than expected.

When the assumption was made that a supernova was involved, many things fell very nicely into place. Galama thought it was easiest to start with the supposition that the supernova followed the same fading behavior as SN 1998bw. The big difference was that SN 1998bw had exploded at a distance of only 140 million light years away, while the gamma ray burst on February 28th, 1997, took place in a very distant galaxy. To take this into account, the distance of the gamma ray burst first had to be known. At first, astronomers had not been able to determine accurately the redshift of GRB 970228, but George Djorgovski from Caltech succeeded in doing just that in early 1999. With the help of the Keck telescope, he recorded the spectrum of the faint host galaxy and he found a redshift of 0.695, which corresponded to a distance of about 6 billion light years.

By considering how SN 1998bw would have looked at a distance of six billion light years, Galama could fully explain the peculiar light curve of the February burst. Furthermore, the redshift of the light could explain why the afterglow in the beginning of March was so remarkably red in color. In a comprehensive article that appeared in the summer of 2000 in the *Astrophysical Journal* Galama wrote: 'Together with the evidence for GRB 980326 and GRB 980425, this gives further support for the idea that at least some GRBs are associated with a possibly rare type of supernova.'

Other astronomers reacted equally enthusiastically. 'Three cases are enough to convince me,' says Stan Woolsey, adding that he has a 'theoretical prejudice'. In his review article in *Science* of October 22, 1999, Jan van Paradijs wrote: 'The supernova connection provides a direct link between GRBs and crucial events in the evolution of massive stars and galaxies. Using this framework to describe gamma ray bursts promises to lead to a rapid increase in our understanding of this energetic phenomenon.'

Galama's *Astrophysical Journal* article about the possible supernova contribution to the light curve of the afterglow of GRB 970228 served as the last chapter of his dissertation, 'Gamma-ray burst afterglows'. He graduated on December 8, 1999, *cum laude* from the University of Amsterdam, but by then he had already been working for a few months at Caltech in close association with Kulkarni, Djorgovski, Metzger and Bloom, his former competitors.

He is still matter-of-fact and modest. In their small Pasadena house with their own orange tree in the garden, Anna, whom he married in May 1999, said that when a city official asked Titus where he worked, he had simply answered, 'at Caltech'. 'If they had asked me what kind of work my husband does', she says with the warmth of her Iberian nature, 'I would answer proudly that he is a doctor of astrophysics and he has a three-year position as a Fairchild fellow at the California Institute of Technology.'

And the Anna correlation? The Fairchild fellow and his Spanish bride don't worry themselves about that anymore. Since the number of afterglows of gamma ray bursts have increased, they obey the laws of statistics a little better. Today, they appear with equal frequency on all the days of the week, and that is as it should be.

11 Alchemists of the cosmos

'The time has come', the walrus said
'To talk of many things.'

In November 1572, a new star sullied the immaculate skies above the gently sloping hills of Denmark. Five years later the crystalline heavens were pierced by a comet. On the eve of the discovery of the telescope, astronomy was wresting itself from the iron grip of classical Greek thought. And the eccentric astronomer Tycho Brahe represented the pivotal point in the history of astronomy.

Tycho (as a nobleman, he was called by his first name) had become famous at home and abroad because of his research on the *Stella Nova* of 1572. To keep the young astronomer from being lured to foreign lands, the Danish king Frederik II gave him as a present in 1576 the island Ven, a strategically situated limestone rock in the Oresund between the Danish capital of Copenhagen and the Swedish Landskrona. He built himself a posh Renaissance castle on the highest hilltop of the island, as though a few meters above ground level would help him better to study the stars. He named it Uraniborg. And nearby, he also built an observatory which lay half underground; this he called Stjerneborg. It was equipped with gigantic measuring instruments to help him accurately fix the positions of the planets and the stars.

Tycho did not live long enough to see the discovery of the telescope (he died in 1601 at the age of 54) but his observations were so precise that his assistant, Johannes Kepler, was able to derive the elliptical orbits of the planets. It was on the island of Ven that the fundamentals of modern astronomy were laid down.

All this happened against the will of the local farmers. Tycho was so hated by the island dwellers that within a year of his leaving in 1597, Uraniborg and Stjerneborg were leveled to the ground. He ruled like a potentate; he treated the farmers as slaves and those who refused to work for him, he locked up in a dungeon.

Tycho was a deeply religious man who not only studied astronomy but continued to steep himself in astrology, alchemy and numerology. In the cellar of Uraniborg there was a complete laboratory, with sixteen chemical ovens where mercury and sulphur were mixed to make gold. Tycho designed horoscopes for the sons of the king and he completely based the ground plan of his palace and its botanical gardens on mystical, magical numerology.

In addition to having his own forge, Tycho also had his own printing press, one of the first in Scandinavia. And to be sure he had enough paper, he built a water powered paper mill in 1590 on the southwest point of the island. This required setting up a complete irrigation system because there are no rivers on Ven.

With his golden nose – a prosthesis he had had made because his own was ripped away in a duel – and with his compulsive behavior and intense absorption in things mystical, he must have presented an eccentric picture indeed. But what a meticulous observer he was! The bright comet of 1577 which Tycho had seen from Stjerneborg was also observed at exactly the same place between the stars in Prague. Therefore, it could not be an atmospheric phenomenon as Aristotle had claimed some two millennia before. The comet had to be far beyond the orbit of the moon and it set its own course through the cosmos. This proved that the sun, the moon and the planets were moving freely rather than in transparent, crystalline spheres, otherwise these would have been mercilessly shattered by the comet.

Tycho was no longer surprised by new revelations. Similar observations of the new star that appeared in the constellation Cassiopeia in November 1572 made it clear that this *Stella Nova* was also further away than the moon. From one night to the next, figuratively speaking, the classical image of a god-like perfection and an unchanging sky was ready for the scrap heap. The new star was a blight on the immutable cosmos of the old Greeks.

Tycho measured the position of the newcomer, registered its

luminosity, and observed the star for several months as it became fainter and fainter until it totally disappeared from sight. To measure is to know, but the observations by the Dane told nothing about the true nature of the *Stella Nova*. Was this really the birth of a new star? Or was it a mystical sign from heaven?

Tycho's student, Johannes Kepler, was convinced that the last option was true. Kepler was a sickly, stammering mathematician who believed, as did his master, in numerology and astrology; it is told that his mother was almost burned at the stake for witchcraft. In September 1604 another *Stella Nova* appeared, this time in the constellation of Ophiuchus, the Serpent Bearer. It was seen shortly after a remarkable conjunction of the planets Mars, Jupiter and Saturn had taken place in the same part of the sky. It was Kepler's opinion that the new star was a sign from heaven 'caused' by the three planets coming together at the same time. With the mathematical formulas that he had devised to chart the motion of the planets, he figured that in the year 6 BC a similar conjunction had occurred. If that historical conjunction had also caused a *Stella Nova*, then that could have been the miraculous Star of Bethlehem.

Tycho Brahe and Johannes Kepler did not know that it wasn't the birth of a new star that they had witnessed, but rather, its violent death. It was a star many times more massive than our own sun that spattered apart in a fraction of a second. The outer mantle of the star was hurled with incredible velocity into space while the compact core collapsed leaving behind its mortal remains, which were barely measurable and had very bizarre properties.

About 300 years later, in 1932, the Swedish astronomer Knut Lundmark first used the term 'supernova' to describe such a phenomenon; a term that the American astronomers Walter Baade and Fritz Zwicky brought into use in their theoretical work. Alas, in all of these three centuries there have been no new bright supernovae observed in our own Milky Way galaxy, nor in the seven decades since Lundmark's publication. That there were two such occurrences within 32 years in the time of Tycho and Kepler is sheer coincidence.

Engraving of Tycho Brahe studying a supernova explosion in the constellation Cassiopeia, which took place in November 1572.

The nobleman and his student hadn't the foggiest idea about the life of a star. Even Baade and Zwicky who developed their supernova theories in 1934 were nonplused as to where stars got their energy. Only in the second half of the twentieth century did a sufficiently coherent picture of cosmic evolution emerge to motivate astronomers to explore the mystery of the birth, the life and the death of stars and to discover how indispensable supernovae are for life on earth.

When the universe was born some 14 billion years ago, it was composed primarily of the two lightest elements: hydrogen and helium. These are still the most abundant; heavier elements such as carbon, oxygen, iron and gold are very rare cosmologically speaking; they just 'pollute' the ocean of hydrogen and helium.

Stars too are made up, roughly, of 75% hydrogen and slightly less than 25% helium. They are colossal spheres of gas that are born from contracting clouds of gas. As a result of gravity, a small amount of condensation in a thin interstellar gas cloud builds up and soon collapses under its own weight. This, in turn, makes the density and temperature rise quickly. The protostar, which is often still enveloped in the cloud of dust and gas from which it came, begins to send out heat radiation. It is this radiation that can be detected with an infrared telescope.

If nature itself did not try to interfere, the gravitational collapse would end in the formation of a black hole, that phenomenon where matter is so violently and so unremittingly pressed together that not even light can escape. The star would die before it even established itself in the world – a stillborn child.

Fortunately, gravity does not control everything all of the time in nature. Under the influence of the strong nuclear force deep within the newly formed star the hydrogen atoms fuse together into helium. What earthly alchemists pursued for centuries, i.e., the transformation of one chemical element into another, the cosmos can do without retorts, flasks or test tubes. When the pressure and the temperature are high enough, nuclear fusion takes place spontaneously. The same forces of nature that make the destructive power of the hydrogen bomb possible are also responsible for the production of energy in stars. With the hydrogen fusion an enormous amount of energy comes free and this nuclear energy offers resistance against a collapse. The star stops shrinking and begins a long and stable phase of its life in which gravity and gas pressure are in equilibrium.

Our own sun has been in this stable phase for about four and a half billion years. In the innermost part of the sun, where the temperature is 15 million degrees and the pressure is 3 billion atmospheres, 570 million tons of hydrogen are converted to helium every second. Seven thousandths of that, which is 4 million tons per second, is immediately converted into energy. That means the sun becomes 4 million tons lighter every second. Fortunately, the supply of fuel is so incredibly large that this cosmic fusion reactor, with a capacity of 400 quintillion megawatts, can keep going for billions of years.

Our own Milky Way galaxy has a good 200 billion stars and the number of galaxies in the observable universe is also in the neighborhood of a couple of hundred billion. The cosmos contains tens of billions of trillions of nuclear fusion power stations like the sun, each of which works against the tyranny of gravity. Every little point of light in the sky marks a long-lasting fight to the finish between the fundamental forces of nature.

Ultimately, however, it is indeed gravity that wins. Gravity needs no fuel. While a star stokes itself up to prevent a catastrophic collapse, gravity just has to sit by patiently and wait. When the hydrogen supply finally gets used up, the internal pressure drops and gravity makes its move.

How fast this occurs depends upon the original mass of the star. A relatively small and low-mass star such as the sun can survive 10 billion years on its supply of hydrogen. Still smaller dwarf stars can reach a lifetime of 100 billion years. But big, massive stars have to keep feeding their nuclear ovens faster and faster to resist the pull of gravity. Although their supply of fuel is larger, they race through it so quickly that they touch bottom after 10 to 100 million years.

But the moment will arrive when the hydrogen supply in the deepest interior of our own sun is used up, a couple of billion years from now, and then gravity takes over, temporarily leading to a collapse accompanied by an increase in temperature and pressure which brings about the burning of the helium core that has formed. However, in a thick shell surrounding this core, hydrogen fusion is still taking place, and as a result, the helium core grows larger and ever more massive. Because of the shell burning, the outer layers of the sun swell up: the sun has now become a giant star.

In this new guise, the sun will have a lower surface temperature than it does now. Because the energy that is produced in the stellar interior is distributed over a much larger surface area, the outside of the sun is cooler and redder. Moreover, the sun, which is now a red giant, is not as stable as it is today. On the surface of this colossal sphere of tenuous gas, gravity is so low that the gas can easily escape into space in the form of a powerful solar wind.

Nor is there internal stability and quiet. In the space of about one billion years, the helium core has become so massive and hot that new spontaneous fusion reactions take place, converting helium into carbon. And still later, carbon atoms fuse into oxygen. These new fusion reactions occur much more rapidly so that in a relatively short time, a core of carbon and oxygen form in the solar interior. This core is surrounded by a shell in which helium fusion takes place, while farther out, shell burning of hydrogen continues to go on.

As a result of this exorbitant energy production, the sun swells up even further, until it becomes a red supergiant, large enough to engulf the two innermost planets, Mercury and Venus, leaving nothing of the earth but a burned out, dried up clump of stone. So traumatic are all of the changes that occur in its interior, that the sun goes into a state of disequilibrium. Its outer layers begin to pulsate and in a few tens of thousands of years great quantities of gas are blown into space. A slowly expanding nebula forms around the sun with a remarkably symmetrical structure. Similar nebulae surrounding other stars were first discovered a few centuries ago, and because, at first sight, they resembled dimly lit planetary disks, they were called planetary nebulae, a name that is still in use, even though the gaseous shells have nothing to do with planets.

What is left over from the sun is a relatively small core that is primarily made up of carbon and oxygen surrounded by a thick mantle of helium and a thin outer shell of hydrogen. The fusion reactions in the core come to a stop because the temperature and pressure never get high enough for the fusion of oxygen atoms to get started. Under the influence of gravity the sun collapses into a super-compact sphere of gas which is barely bigger than the earth. The remaining internal heat of this dwarf star is now distributed over a very small area and the surface temperature is so very high that a blue-white light is radiated.

Every star that is about as massive as the sun ends its life as a white dwarf, and there are actually large numbers of them. The only trouble is that they are so faint they are difficult to find. They are also rather bizarre objects in which the gas atoms are rigidly packed one on the other. The mass is comparable to that of the sun but the volume is more like the volume of the earth. One cubic centimeter of white dwarf matter weighs about 1000 kilograms and the gravity on the surface of a white dwarf is ten thousand times as strong as the surface gravity of the earth.

Where does a white dwarf get its energy? Not from nuclear fusion reactions because these have stopped. No, it is from the gradual cooling of the hot dwarf star that permits it to radiate a small amount of light and heat for billions of years more. The radiation that is sent out becomes less and less energetic and finally the white dwarf turns into a dark, cold cinder; the mortal remains of a sun-like star.

What actually keeps the sun from collapsing further? For that we must thank the quantum effects in the interior of the star. Under the extreme conditions on the inside, all the gas atoms are completely ionized: the electrons are no longer bound to the atomic nuclei of which they were once a part. Instead, there is a so-called plasma, consisting of positively charged atomic nuclei and negatively charged electrons that move freely among each other. When this plasma is pressed together by gravity to a density of about one ton per cubic centimeter, further compression is hindered by the quantum properties of the electrons and it stops.

With stars that are considerably more massive than the sun, the process is very different. To begin with, they have a much shorter lifetime because they burn up their fuel supply at a much faster rate. In the nucleus of a star that is twenty times as massive as the sun, the fusion reactions must occur much faster in order to withstand the stronger pull of gravity. The result is that such a star produces almost 50,000 times as much energy as the sun and the supply of hydrogen in the core is used up in just10 million years.

The helium burning phase, during which the star swells up to a supergiant, also lasts for a much shorter time than it does in the sun, about 1 million years. The next stage that prevents the massive star from continuing to collapse is the fusion of carbon. In the fusion of carbon, which lasts for about 10,000 years in a massive star, the elements of magnesium, sodium, neon and oxygen, among others, are made. When the carbon supply in the core is also totally consumed, the star is compressed even more by the enormous force of gravity and the temperature of the nucleus increases to more than 1 billion degrees. Under these extreme conditions neon and oxygen atoms fuse into sulphur, silicon, phosphorus and aluminum, to name a few. This phase does not last more than about10 years. The demise of the star is fast approaching.

In a process that finishes within a week, the temperature of the core increases to 2 or 3 billion degrees and the silicon atoms fuse together to form iron atoms. At that moment the core of the star is composed primarily of iron (not as a solid but in the form of a fully ionized gas) and it is then surrounded by a series of concentric shells in which oxygen, carbon, helium and hydrogen are burned. The hydrogen-rich outer layers of the massive star, with a total mass of perhaps ten solar masses, may have been blown out into space at an earlier stage of its evolution, in which case the final mass of the star has been reduced considerably.

At this point the patience of gravity is rewarded. Once iron nuclei are formed in the core of the star, spontaneous fusion reactions stop because iron is the most stable element in nature and its atoms do not spontaneously fuse together to become heavier atomic nuclei. The internal pressure of the massive star produced during the feverish sequence of the increasing tempo of fusion reactions, drops off sharply and the star collapses totally in on itself by its own weight.

In a fraction of a second the iron core of the star that started out as big as our sun, shrinks to a super compact ball with a diameter of some tens of kilometers. In the process, the iron atoms are completely ripped apart into their constituent protons and neutrons, between which free electrons can still be found. The quantum properties of the electrons are no longer able to counter the immense pressure from the outside as they do in a white dwarf. Instead, the negatively charged electrons are pressed into the positively charged protons and fuse into new neutrons with no electrical charge. And so the nucleus of the star transforms into a 20 to 30 kilometer diameter ball of tightly packed neutrons with an unbelievable density of 100 million tons per cubic centimeter. We now have a neutron star.

The outer layers of the star have, as yet, no idea of what is going on deep inside the star. With ferocious speed the layers plunge inward. The temperatures increase at an alarming rate and everywhere in the star new fusion reactions occur which quickly produce an enormous variety of new elements. This nuclear witches' brew, however, does not exert enough internal pressure to prevent the gravitational collapse of the star. Nevertheless, the implosion of the star is soon turned into an *ex*plosion. Because the inwardly racing gas of the star crashes on the almost incompressible surface of the neutron star, a powerful outward shockwave arises which completely demolishes the star.

An essential ingredient in creating this supernova explosion is the energy that comes from a flood of neutrinos from the core. These are elusive particles that have no electrical charge and practically no mass; they are produced when a neutron star is forming and they move out with the speed of light to dump part of their enormous energy in the outer layers of the dying star. Without the contribution of this neutrino tidal wave there might never be a real stellar explosion.

Also during the explosion, a number of nuclear reactions take place that not only have to do with nuclear fission and nuclear fusion but also with the capturing of individual neutrons to form particularly heavy elements. In this way, the nuclei of the atoms of such elements as copper, tin, lead, platinum and gold come into existence; they are all heavier than those of iron. In addition, there are huge quantities of radioactive elements formed, among which is cobalt-56. From the decay of these radioactive elements, so much energy comes free that within several months the luminosity of the stellar explosion rivals that of an entire galaxy.

A supernova explosion is the death cry of a massive star. The greatest part of the star is ejected, leaving only the core which collapses into a small, super-compact neutron star. The material that has been blown away forms an irregular nebula that becomes ever fainter and finally flies off into interstellar space, like the streamers of smoke from a spectacular fireworks display. In the Milky Way galaxy there are many such supernova remnants found; the best known one is the Crab Nebula in the constellation of Taurus (the Bull), which was formed by a supernova explosion in 1054.

And what about the neutron star that was left behind? This is much harder to find. Neutron stars are very small – barely larger than the city of Amsterdam – and they send out practically no visible light. The one way they betray their presence is through their strong gravitational field or by their very fast rotation. In the first case, the neutron star has to be part of a binary star system. In the second case, it must have a special orientation in relation to the earth.

Only about one in three stars is alone throughout its life. The sun is an example of such a solitary star. All the others in the universe are either double or multiple stars moving around one another in orbit under the influence of their mutual gravity. Whenever such a binary component undergoes a supernova explosion, the resulting neutron star in many cases continues to describe an orbit around its companion, or rather, both stars follow an orbit around a common center of gravity.

Even though the neutron star is too faint to be observed from the earth, its companion star is usually visible. From a periodic shifting of the wavelength of the light from the star, i.e., the doppler effect, astronomers can deduce that it is part of a double star system; and if the mass of the visible star is known, then the mass of the invisible star can be calculated. When it appears that the invisible star is more massive than the sun it must be a neutron star. A normal star with such a mass would be easy to see.

In some cases the indications are even stronger. If the visible companion swells up into a giant star, it cannot hold on to its outer layers of gas. The gas then streams over from the normal star to the neutron star and as a result of the strong gravitational field from the neutron star, the gas eventually comes with extreme speed to the surface. There it becomes very hot and begins to send out high-energy X-rays. Astronomers have discovered many such X-ray binary stars and in most cases there is little doubt that the invisible component of the double star is a neutron star.

Sometimes, solitary neutron stars can also be discovered from the earth, when their rotational axis has a certain orientation in relation to the earth. During the gravitational collapse of the original stellar core, the speed of rotation increases considerably according to the law of the preservation of angular momentum. We are familiar with the same phenomenon in ice skating. When a skater executes a pirouette with outstretched arms and then pulls the arms in, he or she automatically spins around faster. Newly born neutron stars have a rotational speed of many tens, or even hundreds of revolutions per second.

Such a rapidly spinning neutron star is surrounded by extremely powerful electrical and magnetic fields. In this magnetosphere, electrically charged particles (mainly electrons from the neutron star's surface) are accelerated almost to the speed of light and they produce large amounts of radio waves, visible light and X-rays, especially above the magnetic poles. Because the magnetic poles do not usually coincide with the rotation poles, the spinning neutron star is like an out-of-control lighthouse that spews rotating beams of radiation into the universe. If the earth happens to be in the orbit of the sweep of the beam, then with each rotation of the neutron star (which can be as much as hundreds of times per second) a short pulse can be observed.

In 1967 a British graduate student, Jocelyn Bell, was the first to discover such a 'pulsating star'. With a rather primitive radio telescope at Cambridge University, a short but strong radio pulse was observed every 1.3 seconds in the constellation Vulpecula (the Little Fox). The signal repeated itself with such incredible precision that astronomers seriously thought that they were picking up an artificial radio signal from an extraterrestrial civilization. For a while the mysterious pulsating radio source was referred to with the code LGM-1 with LGM standing for 'little green men'.

Since then there have been over a thousand radio pulsars identified, each of which is a neutron star that happen to be oriented towards the earth in a very exceptional way. Some are part of a binary star system, some are also pulsating in X-rays and in visible light and some have a rotational speed of more than 500 revolutions per second. Pulsars serve as a kind of natural research laboratory where astronomers have the opportunity to look in the greatest detail at the aberrant objects that come into existence when a massive star has dramatically given up the ghost in a totally annihilating supernova explosion.

Supernovae play a prominent role in the life cycle of massive stars, but the high-energy explosions also have a great influence on the evolution of the universe as a whole, and are even of critical importance in the development of life. Stars are cosmic alchemists that take the primitive matter that came into existence shortly after the big bang – essentially hydrogen and helium – and via nuclear fusion reactions convert it into a rich mixture of chemical elements such as carbon, magnesium, sodium, neon and oxygen. Through stellar winds and the formation of planetary nebulae, some of the elements produced by fusion come into the space between the stars, but these are in relatively small amounts. Supernovae, in contrast, produce not only a much richer cocktail of heavy atomic nuclei, they are also the biggest 'polluters' of the universe and have, in the course of cosmic history, enriched interstellar space with almost every element in the periodic table.

The earth upon which we live – its core of iron and nickel, its mantle of silicates, its radioactive uranium ore and its rare veins of gold – consists mainly of the cast-off products of massive stars, blown into space by powerful supernovae explosions. The glass and aluminum that are used in making the giant telescopes with which astronomers search and re-search the universe, would be non-existent were it not for supernova explosions. Nor would Tycho Brahe have had his gold nose, which he might have pensively rubbed when he left his alchemy laboratory in November of 1572 to look in wonderment at the unannounced *Stella Nova*. He owed thanks to similar sorts of *Stella Novae* that succeeded in changing everyday matter into gold and other noble metals some billions of years before.

And what holds for the non-living matter found all around us holds just as well for the living matter from which we are made. The carbon atoms in our muscles, the calcium in our bones, the iron in our blood and the phosphorus in our DNA – it is all stardust, prepared in the nuclear ovens of other suns, spewed out by dying stars and finally coming together in the swirling cloud of gas and dust from which our solar system was born a little less than five billion years ago. Human beings, with all their capriciousness and buffoonery, their intellect and sensitivity, their curiosity and their ignorance, form an impressive but temporary phase in the cosmic cycle that is driven by supernova explosions. And each element plays its own characteristic role and each elementary particle is always used again in a never-ending dance of form and strength, of matter and energy.

Supernovae are the instigators of the evolution of the cosmos. The catastrophic explosions that cause massive suns to be blasted apart, leaving only bizarre super-compact blinking beacons, play a leading role in the complicated drama of the universe. It should not be surprising that they are also responsible for the mightiest explosions since the big bang and that the incredible energy of cosmic gamma ray bursts is so closely related to the astronomical cycle of birth, life and death.

12 The magnetar attraction

At any other time, Alice would have felt surprised at this, but she was far too much excited to be surprised at anything now.

Chryssa Kouveliotou wanted to be an astronaut. In the summertime at her parents' vacation house on the beach, she would lie dreaming her dream, under the fathomless starry Greek sky, that same Greek sky that also inspired such philosophers as Aristotle, Hipparchus and Aristarchus more than 2000 years before. One day her father gave her a beautifully illustrated book about the universe and from that time on, her mind was made up. She would somehow make her way to the stars.

Chyrssa did not become an astronaut but she did discover that she could bring the universe closer without leaving the earth. She would be an astronomer. To experience virtual trips into space with the help of large telescopes became her new dream. At the University of Athens she studied physics; no ordinary choice for a Greek girl. Out of the 120 students there were only 10 women. 'Greece was a cloistered, stifling country', says Chryssa Kouveliotou now. She wanted to move on as quickly as possible.

After four years Kouveliotou left for England where she earned her Master's Degree at the University of Sussex in Brighton. Then she asked for a graduate student position at the famous Max Planck Institute for Extraterrestial Physics in Garching, Germany, near Munich. Everything went along without a hitch until she found out that in Garching there was no department for optical astronomy. There was X-ray astronomy, gamma ray astronomy, cosmic ray research, solar physics, infrared astronomy, all with top researchers – but no optical astronomy.

'I was at the end of my rope,' says Kouveliotou. 'Did I have to go and find other places? Or was it possible that one of these departments would have something for me?' She decided to give it a try. She had long discussions with various research groups and, for one reason or another, the gamma ray group clicked with her and Chryssa Kouveliotou became a gamma ray astronomer.

At the end of the 1970s gamma ray astronomy was essentially in its balloon stage. The group leader, Volker Schönfelder, who later became project scientist of Comptel, one of the gamma ray telescopes on board the Compton Gamma Ray Observatory, would hang instruments under huge stratosphere balloons and 'launch' them from a base at Alice Springs in Australia. Schönfelder's team did research on the gamma radiation of the Vela pulsar, a young pulsar in the constellation Vela (the Sails) that is visible only in the southern hemisphere.

In Garching, Kouveliotou came face to face with gamma rays bursts for the first time. The German astronomers had known for a long time about the existence of the mysterious explosions and their gamma ray telescope may well have registered a few bright bursts. Kouveliotou was given the job of developing computer software that could be used to search the observational data for gamma ray burst detections. 'In a short time I learned everything there was to know about gamma ray bursts,' she says. 'And at that time it wasn't much'.

Unfortunately, her work at Garching did not lead to a PhD thesis. Schönfelder's newest telescope, the forerunner of Comptel, was too heavy to be lifted by the crane at the balloon base in Alice Springs, which meant that it could not be flown in Australia. At the United States balloon base in Palestine, Texas, on the other hand, lifting the colossal instrument would be no problem. But the problem in Texas was that the Vela pulsar could not be seen very well from the United States. Instead, measurements were made of Cygnus X-1, a bright X-ray source in the constellation of Cygnus, the Swan. This was fine, but Chryssa ran into another snag. She could not use the research for her PhD because those measurements were being used as the basis of a thesis by another student who had started at Garching just before her.

What now? Stop with astronomy? Look for another wavelength area of interest? Or look for another thesis subject? By sheer coincidence, it turned out that another gamma ray astronomer at the Max Planck Institute, Dieter Hovestadt, was looking for a new team member. Hovestadt had designed a gamma ray detector for the International Sun–Earth Explorer 3 (ISEE 3), an American–European satellite for research on the interactions between the sun and the earth.

ISEE 3 was launched on August 12th, 1978. It had four gamma ray detectors on board, among which was the scintillator designed by Tom Cline from NASA's Goddard Space Flight Center in Greenbelt. Although a detector had been developed at Heidelberg, they had little expertise in the area of gamma ray astronomy. At this point, Kouveliotou had the advantage; she was already trained in the subject and she was a welcome addition to the Hovestadt team.

Kouveliotou had also done research on gamma radiation of solar flares and had contact with solar physicists all over the world. But her thesis on fast, transient gamma ray phenomena was mainly about gamma ray bursts. And one of these was the remarkable burst that took place on Monday March 5th, 1979, which has already been mentioned in Chapter 1.

The March 5th burst was a particularly strong and long-lasting explosion that was observed by many satellites and space probes. Since the burst was registered at different times, it was possible to deduce its sky position very accurately. Afterwards, it turned out that the burst came from a supernova remnant in the Large Magellanic Cloud, a small neighboring galaxy of our own Milky Way.

At that time most astronomers still assumed that gamma ray bursts were no farther than a couple of hundred light years from the earth. The Large Magellanic Cloud is actually 160,000 light years away. If the March 5th, 1979 burst did come from the Large Magellanic Cloud, it must have been an unusually high-energy explosion.

It was only later that it became evident that the March 5th burst was not extremely far away, but on the contrary, very close by. 'Ordinary' gamma ray bursts come from galaxies at a distance of billions of light years away and compared with these 'cosmological' explosions, the one in the Magellanic Cloud was a meager little bang,

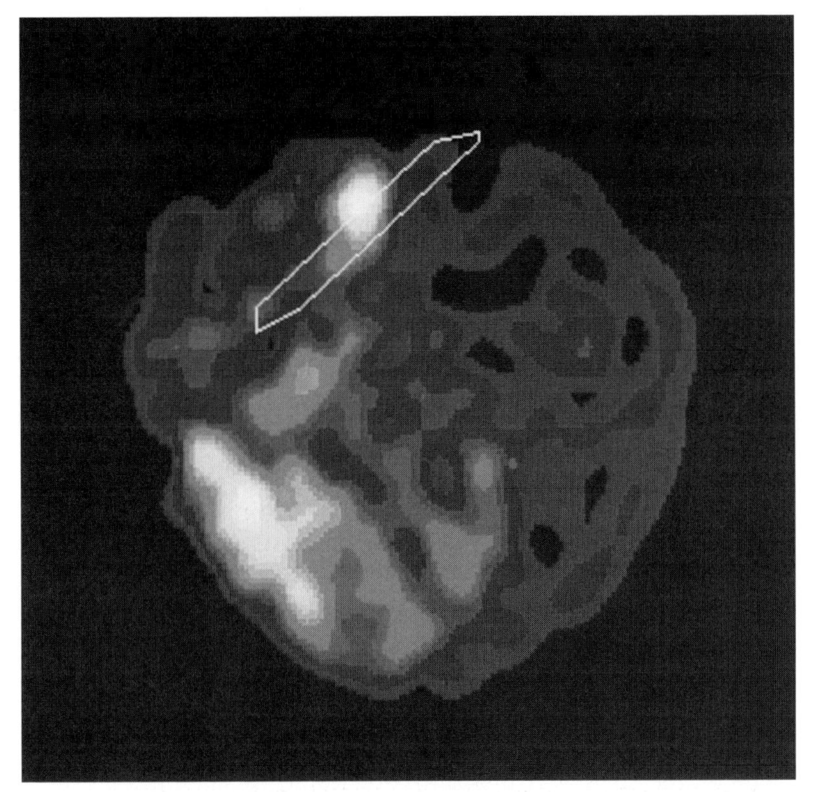

X-ray photograph of supernova N49 in the Large Magellanic Cloud. The white outline marks the error box of the soft gamma repeater SGR 0526–66. If the repeating gamma ray bursts come from a neutron star, it must be far away from the center of the supernova remnant.

even though it sent out more energy in a few minutes than the sun does in 1000 years. Moreover, this was not a one-off phenomenon: in the early 1980s new gamma ray bursts were observed again coming from the same direction.

Because the bursts came from a supernova remnant (named N49) the conclusion was drawn that the explosions take place on the surface of a neutron star. Also, the fact that the gamma radiation which was sent out was much 'softer' (less energetic) than the radiation of a 'regular' gamma ray burst, suggested that a completely different phenomenon was being observed. Even though no one had used the term at that time, astronomers had discovered the first soft gamma repeater (SGR), a new type of object that later brought world fame to Chryssa Kouveliotou.

After receiving her PhD in 1981, Chryssa returned to Greece where she took a position at the University of Athens. 'A stupid decision', she says now. 'Greece was not geared for observational astrophysics. There were no data to work on, Internet connections were not available in the 1980s, and there was not much to do scientifically. It was frustrating.' She regularly made trips to visit her old colleagues at the Max Planck Institute and it was during one of her visits that she met her future husband, Jan van Paradijs, who had come to give a lecture on X-ray binaries. 'We named him the Bird of Paradise', she says. 'He had a strange name and he also looked strange with his short hair, flamboyant shirt, tight pants and a black pearl earring.' Since van Paradijs died, Kouveliotou wears the pearl, which Jan had given her earlier as a birthday present, as a pendant around her neck.

During the 1980s she used every possible excuse not to go back to Greece. She did a two-year stint at NASA's Goddard Space Flight Center, where she studied the observational data from Brian Dennis's group on the Solar Maximum Mission (SMM), a satellite that spent most of its time studying X-rays and gamma rays of solar flares. 'The SMM instruments also registered gamma ray bursts,' recalls Kouveliotou, 'and in the evenings and weekends I spent my time studying them. Putting it all together, I found about 150 gamma ray bursts in the SMM data.'

During Kouveliotou's stay at Greenbelt, Kevin Hurley, who still worked in Toulouse, organized a meeting on gamma ray bursts where he reported on unusual repeating sources of short, soft gamma ray bursts, comparable to the burst of March 5th, 1979. 'These other sources had also been seen for the first time in 1979', says Kouveliotou, 'but it wasn't until 1986 that we realized we were dealing with an entirely different class. I remember all of us asking ourselves if we should name them HXRs or SGRs – hard X-ray repeaters or soft gamma repeaters. We finally chose the latter.'

Unlike the 'normal' gamma ray bursts that are identified with the letters GRB and the date the burst was seen, the soft gamma repeaters were identified by their position in the sky. The first soft gamma repeater that was discovered in the Large Magellanic Cloud now has the official designation SGR 0526–66 because it was located at right ascension 5 hours 26 minutes and at a declination of −66 degrees,[1] in the constellation Dorado (the Goldfish). The two new sources that were being talked about in Toulouse are SGR 1806−20 (in Sagittarius, the Archer) and SGR 1900 + 14 in Aquila (the Eagle).

But giving these strange, repeating sources names and numbers still does not tell anything about the objects themselves. Maybe they were neutron stars: the two new SGRs are in the plane of the Milky Way, where most of the supernova remnants and neutron stars are found, but the kind of phenomenon that would lead to such high energy explosions remained an unanswered question.

In the summer of 1987 Kouveliotou's sabbatical from the University of Athens came to an end and she returned to Greece. But she still went back to Goddard during her next summer vacations. It was there that she attracted the attention of Jerry Fishman who was searching for an instrument specialist for his BATSE detector. 'In Athens I had gotten a permanent position as assistant professor', says Kouveliotou 'but I chose to go to work for 2 years again in the United States.' In January 1991, a few months before the Compton observatory was launched, she came to the Marshall Space Flight Center in Huntsville. 'It was a fantastic time', she says. 'I was completely caught up in gamma ray bursts. I did the greatest part of the data analysis single handed and I knew all the bursts as though they were personal friends.'

Van Paradijs at that time worked very closely with the astronomers at Marshall and shuttled back and forth between Amsterdam and Huntsville. Kouveliotou often flew along with him. 'Delta Airlines made a fortune on us', she says. On October 22, 1992, she married her Bird of Paradise and the following year she gave up her position at the

[1] Right ascension and declination are coordinates on the sky, and compare with geographic latitude and longitude on the surface of the earth.

University of Athens and emigrated permanently to the United States.

In 1995 van Paradijs was offered a part-time professorship at the University of Alabama in Huntsville. 'Ed van den Heuvel [director of the Amsterdam institute] wasn't too happy about that,' says Kouveliotou, 'because Jan would be spending 4 months out of the year in Huntsville. But Amsterdam ultimately got a lot in return; there was an intense and productive network built up between the two research groups.'

Chryssa and Jan became an inseparable duo. They went together to symposia and meetings all over the world, they often worked together on scientific publications, they traveled regularly between Amsterdam and Alabama and rested for a couple of weeks each year at a camping site in sunny Greece. And in the meantime they became more and more specialized in gamma ray bursts and soft gamma repeaters. That finally led to the discovery of the first optical afterglow by the graduate students of van Paradijs and it also led to the name of Kouveliotou being forever bound to the discovery of magnetars.

That was the first great breakthrough in 1998: soft gamma repeaters are magnetars – neutron stars with an incredibly strong magnetic field. The gamma ray explosions come from powerful star quakes, and the enormous magnetic field acts as a brake on the rapidly rotating neutron star. This idea was conceived of and calculated by two American theoreticians but it was Kouveliotou who came up with the proof in early 1998. Just a short time later she discovered the fourth soft gamma repeater. And it was also in that same year that the earth felt the hot breath of a magnetar explosion. 'It was a most remarkable year', according to Kouveliotou.

The magnetar theory was proposed in 1992 by Robert Duncan from the University of Texas in Austin and Christopher Thompson from the University of North Carolina in Chapel Hill. Duncan and Thompson found it surprising that pulsars have a weak magnetic field, relatively speaking; 'only' a few billion gauss (as compared to the strength of the magnetic field on the surface of the earth, which is 0.6 gauss; even the little magnet on your refrigerator door has a magnetic field strength of about 100 gauss).

Naturally, a few billion gauss from an earthly point of view is very large, but, as the two theoreticians calculated, Mother Nature should have no problem creating objects with a field strength a hundredfold greater. Nature's trick lies in the fact that the core of a massive star, at the end of its life during the supernova explosion, completely collapses to a compact neutron star with a diameter no greater than some 20 kilometers. With the gravitational collapse, the star is swept up as a result of the law of conservation of angular momentum. This gives the newly born neutron stars rotational speeds of tens or hundreds of revolutions per second. But because the magnetic field lines are, so to speak, anchored in the stellar gas, the strength of the magnetic field also increases at a frightening rate.

Duncan and Thompson decided to try to find out what such a highly magnetized neutron star, which they called a 'magnetar', would look like. This was no trivial assignment; neutron star matter is not found on earth and it is impossible to simulate in the laboratory. To find out something about it you have to refer to complicated calculations in the border area between solid state physics, relativity theory, quantum mechanics and plasma physics.

Anybody who has ever attended a presentation by these two theoreticians can understand why many people believe that astronomy is a difficult profession. When you first see Duncan, he looks more like a solid tradesman than a theoretical astrophysicist. During a lecture he races through any number of overheads with formulas and equations at such a speed that even his astronomer colleagues can lose track of what's going on. Thompson, on the other hand, with his very high forehead and somewhat 'out of it' demeanor, reinforces the cliche picture of the proverbial absent-minded professor. He is much calmer but he sometimes speaks so softly that the drift of his presentation is lost upon all but the select few.

Be that as it may, the two of them came to the conclusion that a newly formed neutron star could easily have a magnetic field strength of 1 thousand trillion (10^{15}) gauss and such a magnetar would also be observable because of the irregular explosions of X-rays and gamma rays it produced. Soft gamma repeaters are magnetars, according to Duncan and Thompson.

A field strength of a few hundred trillion gauss is really beyond the imagination. If a magnetar happened to be 200,000 kilometers away, about half the distance between the earth and the moon, the magnetic field would be so strong that it would pull your car keys out of your pocket and wipe out your credit cards. And on the compressed surface of the magnetized neutron star, which is made up primarily of iron, the magnetic forces are several hundred million times as strong as gravity is on the surface of the earth.

The effect of such power leads to strong quakes in the metal crust of the neutron star. Just as in earthquakes, there are small shifts which build up tremendous tension. At a particular moment, the tension discharges itself in a massive 'star quake' causing various parts of the crust to rearrange themselves. The seismic vibrations that then occur produce the high-energy plasma waves in the magnetosphere of the neutron star, which in turn emit X-rays and gamma rays.

On paper the magnetar theory was impressive enough, but how do you find out if nature really does behave according to the calculations of Duncan and Thompson? In 1992 no one was even sure that soft gamma repeaters were neutron stars, let alone that they had a superstrong magnetic field.

But then things began to change quickly. In 1994, Shrinivas Kulkarni and Dale Frail published their article about SGR 1806–20, the second soft gamma repeater, in which they showed that this source also coincided with a supernova remnant just as in the repeating gamma ray burst in the Large Magellanic Cloud. Shortly after their discovery, the Japanese ASCA satellite found a weak X-ray source at the place of the explosion that increased enormously in luminosity at the same time as the gamma ray explosions. An X-ray source in the center of a supernova remnant could only be the neutron star that was created during the supernova explosion.

The identification at that moment was grist for the mill for astronomers like Don Lamb who were convinced that 'ordinary' gamma ray bursts were also associated with neutron stars. Meanwhile, everybody acknowledged that gamma ray bursts and soft gamma repeaters were two different entities, but the observations from the ASCA satellite dispelled any doubt that there could be powerful explosions of high-energy radiation on the surface of neutron stars.

Such arguments did not move Chryssa Kouveliotou. 'The idea that gamma ray bursts and soft gamma repeaters had something to do with each other is ridiculous', she says. 'I have seen thousands of gamma ray bursts and I know SGRs inside out. This has to do with a totally different kind of phenomenon.' In 1997, when the distance of a gamma ray burst was fixed for the first time, it seemed undeniable that she was right. Others could steep themselves in the study of cosmic gamma rays bursts that are billions of light years away in the universe, but for Kouveliotou, her total concentration was devoted to researching the less energetic but no less mysterious soft gamma repeaters.

How could the validity of the magnetar theory of Duncan and Thompson possibly be demonstrated? For Kouveliotou it was as clear as glass. If soft gamma repeaters were actually magnetars, i.e., neutron stars with an extremely strong magnetic field, then it had to do with rapidly rotating objects. Moreover, the theory predicts that the speed of rotation of the neutron star would be slowed down by the magnetic field. Who knows whether or not the rotation could be observed in one way or another, and, if you could watch the suspected object for a long enough time, would it be possible to get on the track of the braking mechanism exerted by the magnetic field?

The first proof that it had to do with rapidly rotating objects was already available on March 5th, 1979: the powerful gamma ray explosion from SGR 0526–66 in the Large Magellanic Cloud showed a notable periodic flickering at an interval of about 8 seconds. Could that be the fingerprint of a rotating neutron star? The supernova remnant in which the explosion took place, N49, is no older than a few thousand years, which means that the neutron star in the center is still very young. But newborn neutron stars rotate much faster than one revolution in 8 seconds. Only a very strong magnetic field can account for this slow period at such a young age. This fits in beautifully with the magnetar theory. In a couple of thousand years the neutron star would be slowed down by the strong magnetic field to such an extent that it would only make eight revolutions per second on its axis.

It would have been good if Kouveliotou could measure the flickering frequency of SGR 0526–66 again to see if there was indeed a small reduction in speed. But it was not as simple as all that. For a long time the object had remained quiet and at the position of the gamma explosions there was nothing to be seen on other wavelengths . Then too, the BATSE detector was not able to make timing measurements precisely enough to permit a minute change in the frequency of the flickering to be detected.

But with SGR 1806–20 the situation was different. The gamma ray source was also observed in X-rays and if the X-rays also revealed the periodic flickering then it could be very accurately measured with the American Rossi X-ray Timing Explorer (RXTE), an X-ray satellite with an enormously high time resolution and sensitivity. Kouveliotou was granted observing time on the RXTE satellite. As soon as SGR 1806–20 registered a new gamma ray explosion, the X-ray detectors would be focused on the soft gamma repeater.

In the beginning of November 1996 it happened. BATSE registered a new explosion (eventually it turned out that there was a whole series of gamma ray explosions) and between the 5th and the 18th of November X-rays from SGR 1806–20 were detected for a couple of hours. Also, in the weeks that followed, the puzzling object in Sagittarius was regularly observed by Tod Strohmayer from the Goddard Space Flight Center, who had requested time on the Rossi satellite just as Kouveliotou had.

The observations were made but it was far from easy to discover a periodic signal in the X-ray measurements. If it existed, it must be so terribly weak that it could only be brought to light by the most detailed analysis. The analysis took months and was done for the most part by Stefan Dieters, a colleague of van Paradijs at the University of Alabama in Huntsville. His work was not in vain. Dieters discovered that the observations by both Kouveliotou and Strohmayer revealed a weak pulsating signal with a period of 7.47 pulses per second, almost exactly the same as the period in the gamma radiation from SGR 0526–66.

If the magnetar theory were correct, the pulsation period of the X-ray source would, over time, slowly but surely increase. It was believed that it would take a few more years before that effect would be measurable. But this is a real case of 'seeing is believing'. Kouveliotou and her colleagues didn't have to wait that long at all! There were X-ray observations from SGR 1806–20 available from 1993 when the X-rays from the soft gamma repeaters had first been detected with the ASCA satellite. At that time there were no X-ray pulses discovered because no one knew on which frequency they might show up. Now that the pulsation period was known, it was possible that the weak pulsing effect could also be found in the old observations.

In the beginning of 1998 the search efforts paid off. A careful comparison of the measurements from 1993 and 1996 showed that the rotation period of the neutron star had increased by 0.008 seconds, a huge amount if you remember that we are dealing with a star that is some 1.5 times as massive as the sun and is not easily braked. Judging from the degree that the neutron star had been slowed down, Kouveliotou figured that the magnetic field had a strength of 800 trillion gauss, precisely in accordance with the calculations of Duncan and Thompson.

'When we were sure we were right, I called Robert and Chris', says Kouveliotou. 'Chris was wildly enthusiastic. "8 times 10 to the 14th gauss!" he kept repeating. Robert was discreetly emotional about it. "Thank you, thank you", he kept saying. Later I heard that he and his wife opened a bottle of champagne.' Very quickly an article was written up for *Nature* which appeared in the May 21, 1998 issue. Magnetars were now placed for all time alongside of the other cosmic rarities.

As if Nature herself wanted to celebrate the discovery of magnetars,

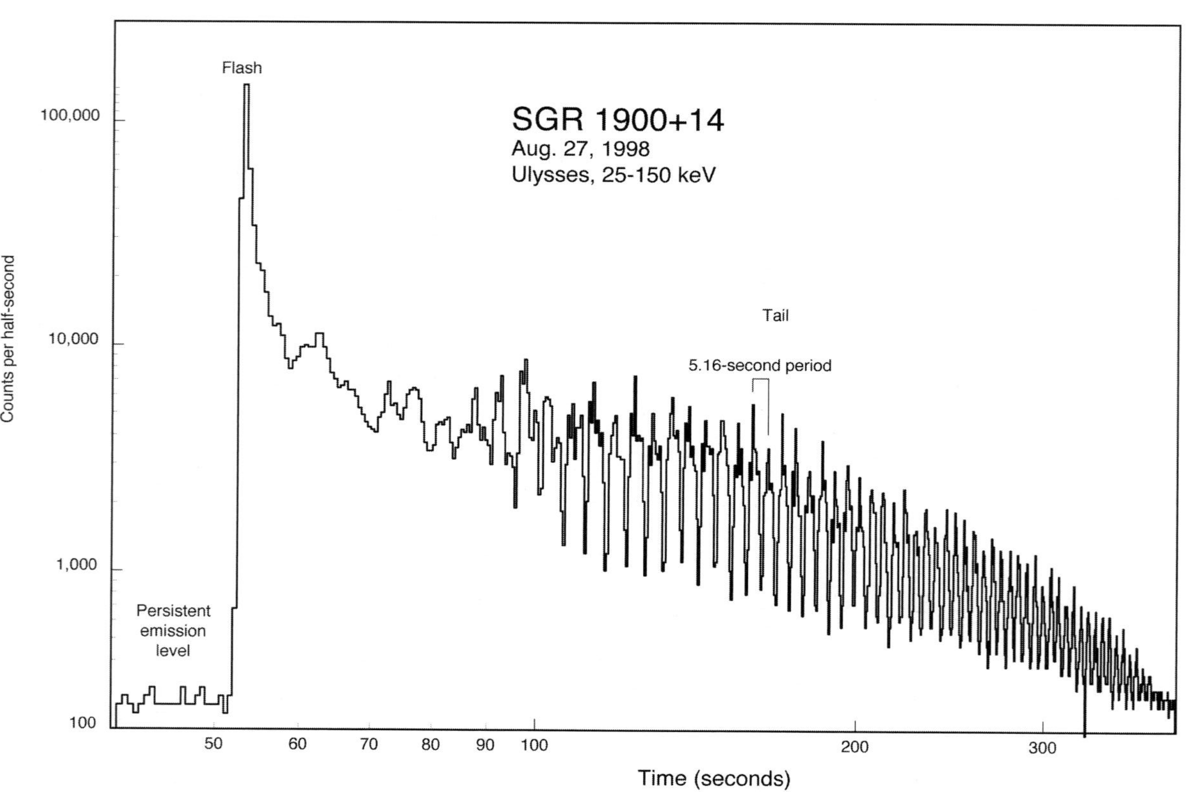

The light curve of the explosion from SGR1900+14 on August 27, 1998. The rapid fluctuations in the strength of the signal are due to the rotation of the neutron star.

a new one flamed up on June 15th in the constellation Scorpio. In one weeks' time BATSE registered no fewer than twenty-six explosions, twelve of which took place on the 18th of June. Thanks to the Inter-Planetary Network the position in the sky of the new soft gamma repeater could be fixed; SGR 1627–41, as it was named, came exactly from the same point in the sky where a supernova remnant in the Milky Way had been observed

But the biggest surprise of the 'magnetar year' 1998 came on Thursday August 27th when SGR 1900+14 in the constellation Aquila, registered a super heavy explosion. Kouveliotou and her colleagues had already been watching the object for a while and had also succeeded, with the help of the Rossi X-ray Timing Explorer, to confirm that this too was a neutron star with a rotation period of several seconds which was slowed down by a very strong magnetic field of about 1 trillion gauss.

On August 27th at 09:22 Universal Time the Rossi satellite was not aimed at SGR 1900+14. A totally different X-ray source was being observed and the casing of the X-ray detector was supposed to keep out radiation from other directions. But the enormous flood of energetic X-rays and gamma rays from SGR 1900+14 that raced through the solar system on that Thursday was so powerful that it was registered by the Rossi detectors. At once the instrument began to go wild; such a powerful signal had never before been observed.

Besides the RXTE, the super-explosion was also detected by several other satellites that happened to be on the right side of the earth, as well as by two space probes that were a long way from the earth. These were Ulysses, in the neighborhood of Jupiter's orbit, and NEAR–Shoemaker, which was on its way to encounter the asteroid Eros. From the difference in arrival time of the burst from SGR 1900+14, its sky position could be accurately fixed. And because the measurements were taken at several wavelengths, it was also possible to determine very precisely the total energy of the explosion. Taking into consideration the distance of SGR 1900+14 of about 20,000 light years, the conclusion was reached that the star quake caused 1 billion, billion times as much energy to be released as the yearly energy consumption of the entire world population.

The gamma ray burst also had an influence on the earth's atmosphere, in spite of its tremendous distance. For the ionosphere (a high, thin layer in the atmosphere), it seemed to have become bright daylight in the middle of the night. The ionosphere owes its name to the fact that the gas atoms are easily ionized when they are bombarded by energetic particles from the sun. The atoms then lose one or more electrons and a thin plasma of electrically charged ions and electrons comes into existence. Long distance radio connections make good use of the reflection properties of the ionosphere.

Normally, ionization only occurs on the day side of the atmosphere. Above the night hemisphere of the earth, outside of the direct influence of the sun, the ions and the electrons recombine and nothing more remains of the amazing electrical properties of the ionosphere. But during the burst of August 27th, while it was still night time in California, researchers at Stanford University registered a sudden increase in the degree of ionization. The enormous amount of energy that was dumped in the earth's atmosphere by the distant magnetar had almost as strong an effect as the energy of the sun, which is at a distance of only 150 million kilometers.

Magnetars are truly bizarre objects – super-compact spheres of tightly packed neutrons, one on top of another, with a cracking, shearing crust of iron, an inconceivably strong magnetic field and gigantic explosions of X-rays and gamma rays on the surface. Still, they are probably less rare than you would be led to believe. Kouveliotou estimates that the total number of magnetars in the Milky Way galaxy must number at least one million. 'We only see them during a very short period of their life', she says.

While a 'normal' supernova explosion produces a 'normal' neutron star that in certain cases can be seen as a radio pulsar, it is estimated that in one case out of ten a magnetar is formed – a super-fast rotating neutron star with an extremely strong magnetic field. According to the theory of Duncan and Thompson, this occurs only when the original

An artist's impression of a magnetar: a small, super-compact neutron star with an extremely strong magnetic field.

star is already rotating rapidly and has a strong magnetic field. For about 10,000 years – just a cosmological blink of an eye – such a magnetar is visible as a soft gamma repeater. The speed of rotation and the strength of the magnetic field decreases and after a few thousand years there are fewer and fewer strong star quakes. Finally, all that is left is a sluggish neutron star that rotates about once every ten seconds. There are no more catastrophic explosions and the magnetar can only be detected by the X-ray pulsations it sends out.

According to Kouveliotou and van Paradijs, older magnetars are also observed in the Milky Way galaxy. Over the past few years astronomers have discovered six 'anomalous X-ray pulsars', or AXPs, having the same properties that fit the predictions of the magnetar theory. With the passage of another 10,000 years, the X-ray pulsations also fade out and the magnetar, which was once so energetic, changes into a cold, dark and torpidly rotating neutron star that cannot be detected from the earth.

The four soft gamma repeaters that Kouveliotou observed have been loosely associated with supernova remnants. They are young magnetars that were formed by the stellar explosion which also produced the nebula. But there are many more known supernova remnants in which there are no observable radio pulsars. Might all of these be hiding an old, snuffed-out magnetar? At what distance is the closest magnetar? Could a star quake just a small distance from the earth be fatal for life on our planet? These are questions that Kouveliotou still does not have the answers to.

When little Chryssa was down at the beach during summer vacation looking up at the stars in the Grecian night sky and dreaming of being an astronaut, she could not have imagined that the unobservable universe was more explosive and more mysterious than the visible and seemingly placid world of stars that the old Greeks had studied thousands of years ago. Her virtual trip to the stars carried her through the cosmological *Guinness Book of Records* and brought her eye to eye with spinning neutron stars, super-strong magnetic fields, repeating gamma ray explosions and devastating star quakes.

The discovery of magnetars is viewed as one of the most important breakthroughs in high-energy physics of the past few years. This turbulent cosmic violence is hidden from the biggest telescopes on earth as well as the lowly human eye, but the X-ray and gamma ray detectors in orbit around the earth bring the invisible into view and the unimaginable to life. Those who would want to discover something new must look to those places where no one has yet looked, or put on glasses that no one has yet worn.

Above all, to really be part of the action one has to get there fast. Five months after the super-explosion of SGR 1900+14 a new revolution was unleashed by a camera that is smaller than the average amateur telescope but reacts faster than a fire alarm.

13 The Argus eyes of Livermore

'You may look in front of you, and on both sides, if you like,' said the Sheep; 'but you can't look all around you – unless you've got eyes at the back of your head'.

Hye-Sook Park does not like to think back to January 23, 1999. 'I had an ache in my heart for two long weeks', she says. 'It was a very difficult period.' For years she had put all her energy into searching for optical bursts but the pleasure of being the 'first' was not to be hers. This fell to Carl Akerlof, who had been her closest collaborator but ended up being her greatest competitor. He was the first to discover the visible light burst from a gamma ray explosion. Why? Simply because it was raining in Livermore on the 23rd of January, 1999.

Almost exactly one year later it is dry and sunny in California. Area 300 of the Lawrence Livermore National Laboratory, east of San Francisco, lies barren and abandoned on a sloping hilly landscape, sparsely overgrown, accented here and there by a formidable cement bunker. When explosion tests are going on, the heavily guarded military terrain is out of bounds for Park too, but today all she has to do is wave her Livermore badge.

On the highest point of Area 300, next to a little stone building, there is a small white container. A tiny room in the little building (that once served as an observation post) barely has room for a couple of computers, a server, an accurate digital clock and a weather station. In the white container sits the apple of Park's eye, LOTIS – the Livermore Optical Transient Imaging System.

'It's been months since I was here last', says Park. 'LOTIS is completely computerized. Each night it goes on automatically.' As soon as it gets dark the cover of the container opens and four sensitive electronic cameras with gigantic telelenses begin to photograph the whole sky. All images are stored in digital form; together they comprise an inexhaustible archive from which astronomers can later look for variable objects, short-lasting phenomena and moving heavenly bodies.

But now and then the cosmic patrol work is suddenly interrupted when the message comes across on the Internet that a new gamma ray burst has been discovered. In a few seconds the cameras are aimed at the indicated part of the sky and they begin to click busily away in the hope that an optical counterpart of the gamma ray burst will be discovered – not a faint afterglow, but a short burst of light that comes at the same time as the gamma ray explosion. Within 10 seconds of the original detection by the Compton Gamma Ray Observatory, LOTIS has already made the first image before any human even knows that there is a new gamma ray burst.

Hye-Sook Park was born in Seoul, but with her interest in the physical sciences South Korea was not the place for her. 'Experimental physics is not popular and especially not if you are a woman', she says. 'Some Korean women still don't understand me at all. "Your husband makes enough money", they say. They just don't understand.'

When Park was twenty she emigrated to the United States and in 1985 she earned her PhD at the University of Michigan in Ann Arbor. Her thesis had to do with the IMB experiment – a large particle detector in an abandoned salt mine under lake Erie. It was designed and built by the University of California at Irvine (I), the University of Michigan (M), and Brookhaven National Laboratory (B). The experiment was originally directed toward research on the possible decay of protons, but in the beginning of 1987 it became front page news with the discovery of neutrinos that came from supernova 1987A in the Large Magellanic Cloud.

It was an exciting period, says Park, but high-energy physics did not offer her an attractive future. 'As a postdoc you quickly land up in one of the big laboratories like Fermilab or CERN where you work on one small facet of an enormous project. It would certainly be much more exciting to oversee a complete experiment', she says. At the University of California at Berkeley she did research on cosmic rays for a while and

(*Left*) *Hye-Sook Park*, project scientist of the LOTIS experiment at the Lawrence Livermore National Laboratory in California.

(*Right*) *Carl Akerlof* is project scientist of the ROTSE experiment which detected for the first time the optical counterpart of a gamma ray burst in January 1999.

in 1987 she got a job at the nearby Lawrence Livermore National Laboratory.

In the mid-1980s Carl Akerlof was having the same sort of thoughts. There must be more to life than particle physics. After finishing his doctorate at Cornell University in Ithaca in 1967, Akerlof worked as a particle physicist at the University of Michigan but the field had become so big that it was no longer attractive. 'What I really needed was to come home at night and feel that I knew something more than I did when I left that morning', he says.

It was the same IMB project on which Hye-Sook had written her thesis that supplied the inspiration for the dramatic change in Akerlof's career. He was involved in a series of remarkable discoveries of highly energetic gamma rays from active galactic nuclei, quasars and pulsars. 'It was a world of difference', he says. 'In particle physics you work with the so-called Standard Model that is damned good in explaining most observations, but in a certain sense it can be dull. High-energy astrophysics is a different story. The field is much less exact, and sometimes it's hard to arrive at a meaningful confrontation between theory and experiment.'

It was around this time that Akerlof heard about gamma ray bursts for the first time but he didn't pay much attention. It was a messy area, he thought, where there was little agreement between the scanty observations and the rather primitive theories. Moreover, he and his colleague Trevor Weekes were too busy doing research at the Whipple Observatory in Tucson on gamma rays and high-energy charged particles in the universe.

But as much as he enjoyed his work, the enormous amount of new observational data that was provided by the Compton Gamma Ray Observatory in 1991 couldn't be ignored. And so Akerlof decided to attend the first Huntsville Gamma Ray Burst Symposium to find out more about it. It was a controversial and contentious meeting. The BATSE observations did not abide by the conventional wisdom that declared gamma ray bursts come from nearby neutron stars, and many of those at the meeting thought the cosmological alternative was too ridiculous for words. 'I came across people who had worked on the same subject for years but feelings were running so high that the air was charged with contention and acrimony', says Akerlof.

It was actually interesting to see a new field of research about such a fascinating problem being mired down in utter confusion. Akerlof realized that there was a real possibility here of contributing to the solution of a critical astrophysical riddle. But how? Not by developing a satellite experiment; such a project would take years with the same disadvantages he found in high-energy physics – too big and too open-ended. No, it had to be a small, comprehensive program and it would have to be done with instruments on the ground.

The idea of looking for optical counterparts of gamma ray bursts was, of course, not new. Since the first days of the Inter-Planetary

Network (IPN) the sky positions of gamma ray bursts had been carefully scrutinized. Unfortunately, it always happened after the fact – a lot of time would pass before the satellite observations were analyzed and a trustworthy error box was derived. The 'posthumous' search for optical counterparts was always fruitless but the one definite conclusion you could draw from this was that optical afterglows from gamma ray bursts were very rare or very dim (in 1997 it became apparent that both descriptions are true). However, what was happening *during* a gamma ray burst remained unknown. It could be that a gamma ray burst might be accompanied by a very short, strong explosion of visible light.

Bradley Schaefer, who is now at the University of Texas in Austin, had already carried out an intensive search for such optical bursts 10 years before. In doing research for his doctorate at the Massachusetts Institute of Technology (MIT) in Cambridge he explored their huge archive of old photographic plates. Perhaps it was possible that one of the old photographs of the sky might happen to show just such an optical burst. Schaefer selected a few small IPN error boxes and looked minutely at each photograph containing that particular piece of the sky, assuming that the mystery object in the error box might produce a number of gamma ray bursts (and hopefully, optical flashes) over the years.

He spent many months in the lovely old library of the Harvard-Smithsonian Center for Astrophysics. 'I must have searched through at least 50,000 photographic plates with a magnifying glass for two years,' he says. 'Sight in my right eye has definitely deteriorated.' At the end of 1981 he found something. On a photo from 1928 he detected a faint little star in the constellation Sculptor that didn't belong there. Photographs of the same area taken 45 minutes earlier and 45 minutes later showed nothing unusual, but on the fourth picture in the series, out of all the stars there was one too many, and this was in the exact position at which a powerful gamma ray burst was observed on November 19, 1978.

Schaefer didn't let himself go overboard. Meticulously he looked for all possible explanations to account for the 'temporary' star, but in the end he concluded that this really had to do with a short optical burst that, in any case, had to come from outside the solar system. In December 1981, he published his discovery in *Nature* and for a long time the tiny star on the photograph from 1928 was considered to be a serious candidate for the optical counterpart of a gamma ray burst.

Since then, it has been established that the puzzling optical burst in 1928 is absolutely not related to the gamma ray burst that occurred 50 years later. GRB 781119 was a 'classic' gamma ray burst and definitely not a soft gamma repeater. Furthermore, the discovery by the Italian–Dutch BeppoSAX satellite indisputably showed that ordinary gamma ray bursts happen billions of light years away, and they are so energetic that they must cause the complete destruction of the exploding celestial body. 'Whatever it was that happened in 1928 still remains a riddle,' according to Schaefer.

Akerlof didn't want riddles, he wanted solutions. Yes, he also was searching for optical bursts, but he wasn't going to look through old archives, he was going to look through a telescope. If you really want to know if a gamma ray burst coincides with an explosion in visible light, you must observe the position of the burst at the moment of the explosion. The best way of course, would be if you could keep the sky under continual surveillance.

Something like that was done on a modest scale by George Ricker at MIT. On Kitt Peak in Arizona, Ricker had installed an array of automatic cameras that were equipped with electronic detectors. This Explosive Transient Camera (ETC) was, in fact, not much more than an array of 35 mm cameras with wide-angle lenses. The only difference was that the photographic films were replaced by charge coupled devices (CCDs), light-sensitive chips like those used now in modern digital cameras. Ricker's ETC recorded everything brighter than magnitude 7.8 (five times more faint than you can just about see with the naked eye on a clear night) but he was never able to detect an optical burst.

According to Akerlof, it shouldn't be hard to do better than a 35 mm camera. Larger lenses with a diameter of 20 centimeters should be able to catch much fainter objects, and if you have enough of them, you can monitor a large part of the sky. At the end of 1991 it looked like such a project might cost several million dollars, and with such a restraint, Akerlof filed his idea in a notebook and let it rest for the time being.

But in the summer of 1992 a unique opportunity presented itself. Akerlof was spending a sabbatical leave at the Center for Particle Astrophysics in Berkeley, collaborating with a group of astronomers led by Charles Alcock from the Lawrence Livermore Laboratory. The group called themselves the MACHO collaboration (for MAssive Compact Halo Object). Their goal was to use CCD cameras in their chase after hypothetical dark objects in the extended halo of the Milky Way galaxy.

In Livermore he met Hye-Sook Park who was also working on the MACHO project. 'I remembered her from the time she was a graduate student at Michigan', says Akerlof, 'and I found it nice to come across someone I knew.' Park and her husband, Richard Bionta, invited him to lunch and during the meal he mentioned that he had heard of somebody at the Livermore laboratory who had built a wide-angle camera as part of the Strategic Defense Initiative (SDI), better known as the Star Wars project. Park reacted enthusiastically. She knew all about it because she had worked on this camera project herself at the end of the 1980s. It had a very unusual 90 millimeter lens with a focal length of 25 centimeters and a field of view of at least 60 degrees.

'The original idea for such a wide-angle camera was to detect enemy missiles approaching from space', Akerlof relates, 'but the project never materialized and the test camera remained unused for a couple of years, stored in an enclosure on one of the parking lots of the lab.' According to Park it was a prototype that was only a quarter of the planned size. It had a complicated optical design (the lens alone had cost 2 million dollars) in which the light that was collected by the lens was channeled to twenty-three separate electronic detectors by means of glass fibers.

Akerlof was very enthusiastic. Such a camera was exactly what he needed. He told Park about his plans to chase down the optical counterparts of gamma ray bursts, and together, in 1993, they presented their idea to Bruce Tarter, the head of the physics department at Livermore at that time. Tarter gave them the green light along with $50,000 to infuse new life into the SDI camera.

This was no small challenge. Together with his graduate student Brian Lee, Akerlof opened the enclosure, brushed the cobwebs away and tried to find out how you were supposed to drive and aim the instrument. The original drive software was no longer available and most of the people who had worked on the project were long since scattered far and wide.

Park and Akerlof worked hard together on the project they dubbed GROCSE, for Gamma Ray Optical Counterpart Search Experiment. They made a great 'odd-couple' team. Hye-Sook Park is a small, feather-light, very intense woman with a sharp tongue and dark eyes that always twinkle with enthusiasm. Carl Akerlof, whose father came from Sweden (the name was originally spelled Åkerlöf), is a large, calm man with somewhat reddish hair and light blue eyes. By nature he has a fatherly quality about him. But for all the disparities of age, culture and background, they worked well together.

Unfortunately, the SDI camera was less well suited to chasing optical bursts than Akerlof had hoped. The unusual wide angle lens had a curved focal surface which was incompatible with flat CCD detectors. To compensate, bundles of glass fibers were needed to conduct the light from the focal plane to image intensifiers and eventually CCDs. But only a small percentage of the incoming photons survived this arduous trip.

Luckily, by the spring of 1993 this last function was no longer necessary, thanks to (of all things) a technical problem the Compton Gamma Ray Observatory was experiencing on board with its tape recorders. Originally, the recorders were used to store temporarily the observational data from the various detectors. Once after each revolution all measuring data were sent to the ground station at the Goddard Space Flight Center. But by the end of 1991, the recorders were showing signs

GROCSE (Gamma Ray Optical Counterpart Search Experiment) was the first automatic camera used to search for optical bursts.

of wear and during 1992 the problem became so bad that the mission engineers had to look for another way of transmitting the data. They chose to bypass the recorders altogether and send the data through to earth in real time via NASA's own Tracking and Data Relay Satellite System (TDRSS).

The good flip side of this procedure was that the gamma ray burst observations from the BATSE detectors were now also available in real time. One second after a gamma ray burst went off somewhere in the universe the data poured into the Goddard Space Flight Center. In principle, that would make it possible to aim a telescope quickly at the right area of the sky. The BATSE positions were not particularly accurate but with a wide-angle lens there would be no problem imaging a potential optical burst if the alert were available fast enough.

During 1992, Scott Barthelmy from the Goddard Space Flight Center started to work on the alert system. The stream of data from the

Compton observatory was examined fully automatically, the observations from BATSE were separated from the others, sudden increases in the number of observed gamma photons were identified and a rough sky position was calculated. The positions could be automatically transmitted to a small automatic camera at the Kitt Peak observatory via the Internet.

Barthelmy's BAtse COordinates DIstribution NEtwork (BACODINE) became operational in July 1993 and Park was the first to knock on his door to ask if she could make use of the data. 'Scott is a fantastic programmer', she says. 'Without him the GROCSE project would never have gotten off the ground.' Park herself put two technicians and a programmer to work on the Livermore telescope to get it to react to the BACODINE alerts and in the summer of 1994 it seemed that everything was in good working order.

The GROCSE project functioned for 2 years, until the middle of 1996. About once a month the telescope responded to a BATSE detection. On none of these occasions was an optical burst discovered but GROCSE was able to set useful upper limits. Because the telescope could detect everything brighter than magnitude 8.5, it was possible to conclude that possible optical bursts must be dimmer than this limiting luminosity.

Even before GROCSE was operational, Park and Akerlof were already working on a new system that would be a hundred times more sensitive at about one percent of the cost. Large format CCD chips were married to commercial telephoto lenses such as those used by sports journalists and the paparazzi. Akerlof, who was himself a dedicated amateur photographer, chose Nikon lenses that cost about $2000 each. He bought three of them.

It soon appeared that this was not the best choice for the purpose. 'We tested the lenses on an optical bench at Livermore', says Akerlof, 'and the picture quality was very disappointing. In the summer of 1994 I borrowed a similar lens made by Canon and it produced pictures of point sources that were twice as sharp!' The Canon lenses were also twice as expensive. As it happened, two of the three Nikon lenses were stolen in the fall of 1994 but fortunately it wasn't all that bad: from the insurance money one Canon lens could be bought right away and three others followed soon after.

Meanwhile, Park worked hard on the camera electronics and the housing. The Livermore lab gave her the freedom to spend a lot of time on GROCSE II, as it came to be known. But all attempts to get support at that time from NASA for the hardware came to nothing.

GROCSE II would have to be more sensitive and faster than GROCSE I. The four telephoto lenses were placed on a single movable platform and together they could cover an area of the sky of 16 × 16 degrees. In a maximum of 3 seconds the instrument could be directed to any desired place in the sky. The CCD cameras were 4000 times as sensitive as the human eye; they could image everything down to magnitude 15.

This could all happen provided, of course, that the financial support would come, but in 1995 that was not yet the case. 'Money was always a problem', says Park. 'Actually the break with Carl comes down to that.' Akerlof saw, with disappointment and regret, how GROCSE II was increasingly becoming a Livermore project. Park devoted more time than anyone else and even Akerlof's graduate student Brian Lee was being paid by Livermore. And when NASA finally came across with $25,000, it also went to Livermore.

Akerlof started to think even further ahead. A 40 cm telescope on a quickly moveable mount could, shortly after a gamma explosion, make a long exposure of the position of the burst and then, in addition to a short optical burst, perhaps it would be possible to discover a fainter afterglow. The Research Corporation was willing to invest in such a project provided the government also put up some money. Maybe NASA would be more interested in supporting a big project rather than GROCSE II.

'At the end of 1995 I talked my ideas over with Hye-Sook', says Akerlof, 'but she was not interested. Instead, she wanted to submit a proposal to NASA to build two camera clusters.' Park in turn, felt intimidated. 'Carl wanted to be more in control', she says, 'while I took

The fully automatic LOTIS camera (Livermore Optical Transient Imaging System) is mounted in Area 300 of the Lawrence Livermore National Laboratory.

care of all the necessary manpower and logistics for the GROCSE project.'

In the beginning of 1996 the relationship between Park and Akerlof deteriorated even more and finally, in March, it all fell apart. Right before the deadline for submitting proposals to NASA, and while Akerlof was just about to leave for a skiing vacation in northern Canada, Park pulled out of the partnership and decided to submit a competitive proposal on her own. 'I was furious', says Akerlof. 'When you come down to it, the whole idea for the project was originally mine.'

Akerlof put in a complaint to the managers of the Livermore laboratory, but for Park, that was going one step too far. 'I have always treated him courteously', she says. 'That sits deep in my Korean character. The name Hye-Sook means "grace". I don't have it in me to offend anyone, but that he should pass me by and go over my head to the management; that was just too much. He called everybody: the head of my department, the head of the physics department, the scientific advisors and even the director of the lab.'

But according to Akerlof, the Livermore staff did all they could to cool things down and bring about some sort of reconciliation. 'Maybe they were afraid of getting a bad name or maybe they didn't want to get the reputation of squelching the ideas of a young, ambitious woman scientist.' However, Park did not get permission to submit her own NASA proposal and Livermore did sign as co-sponsor of Akerlof's proposal. 'At their request, I cancelled my vacation to Canada and flew back to California to face the unpleasant discussions.'

All cooperation came to an end, and unfortunately things never got better between Park and Akerlof. The hardware from GROCSE II was divided between them. Akerlof got the lenses, the camera mount, two CCDs and the weather sensors. Park kept the other CCDs and the actual cameras with all the electronics. The Livermore lab then paid Akerlof $80,000 through the University of Michigan enabling him to buy new cameras and to develop his own electronics.

In addition, Akerlof got $50,000 from the Research Corporation and a NASA grant for half a million dollars divided over three years. A few days after he finished his business with the Lawrence Livermore National Laboratory, he made arrangements with Livermore's competitor, the Los Alamos National Laboratory in New Mexico, where he was able to set up his instruments in 1997.

Meanwhile, Park did continue to get good financial support from Livermore and by the end of 1996 her own camera cluster was ready to be used on a hilltop in Area 300. The Livermore installation was called LOTIS (Livermore Optical Transient Imaging System). At Los Alamos, Akerlof's new instrument became operational by the spring of 1998. It was called ROTSE (Robotic Optical Transient Search Experiment).

It is true that ROTSE and LOTIS look very similar. On every clear night the roof of ROTSE's enclosure opens – ROTSE is housed in an old military communications hut that Akerlof managed to get for $250 from a scrap dealer – and the cameras begin to take pictures of the whole starry sky systematically. As soon as it becomes cloudy, the instrument switches to non-active, and at the first drop of rain the whole enclosure automatically closes. Once every two or three weeks or so, the motors on the camera mount suddenly begin to hum and direct the instrument to face towards that little piece of sky where BATSE, a couple of seconds before, reported seeing a gamma ray burst flare up.

BATSE sees on the average of one gamma ray burst per day, which is about 360 bursts per year. Half of them happen during the day and another 60 during twilight or when there is too much moonlight, when there is no chance to observe an optical burst. From the 120 remaining flashes, half of them are below the horizon for ROTSE and LOTIS and about 20 are too low in the sky to give useful observations. That leaves just 40 bursts, half of which cannot be observed because of cloud cover or other hindering conditions.

By the beginning of 1999 ROTSE and LOTIS had collected hundreds of gigabytes of observational data, but there was not one authentic optical burst, in spite of the fact that they reacted many, many times to burst alerts from the BACODINE network. By then, other astronomers had already found faint optical afterglows but no one could capture the instantaneous visible counterparts of the short-lasting gamma ray bursts.

But on Saturday January 23rd, 1999, all that changed. That morning, at 09h 46m 56s Universal Time, while it was still night-time in North America, BATSE detected a powerful gamma ray burst in the constellation Boötes, just beyond the tip of the tail of Ursa Major. The rough sky coordinates of the burst were distributed via the Internet within a couple of seconds. At Livermore that night, the rain came down in buckets so that LOTIS remained quiescent and useless in its enclosure, but the ROTSE cameras (only three of which were operative at that moment) automatically turned at high speed towards the right direction. Six seconds after the gamma ray burst, the first images were recorded.

Akerlof was looking through his e-mail on that Saturday morning at home in Ann Arbor before eating breakfast and saw that during the night both BATSE and BeppoSAX had observed a gamma ray burst. A quick look at the weather information told him that it had been clear in New Mexico and ROTSE must have made recordings because the constellation Boötes was high above the eastern horizon at that time. After starting the transfer of data files from New Mexico to Michigan he raced to the university to look at the images.

To his horror, the first image was completely black. As a result of a software fault, all the odd-numbered images were lost but luckily, all the even ones were good. The first usable image was taken 22 seconds after the burst. In the meantime, the X-ray telescopes of BeppoSAX had accurately measured the position of the burst and Shri Kulkarni's team from Caltech had also succeeded in detecting a faint optical afterglow.

At around 2 p.m. Akerlof was pretty sure that this time he had a hit. Precisely at the position of the X-ray source and the optical afterglow, the ROTSE images showed a fairly bright star that was not to be found on any of the star charts. On the first image, taken 22 seconds after the burst, the star was hardly brighter than magnitude 12, but on the second image, taken 25 seconds later, the brightness had reached a magnitude of no less than 8.9. If anyone happened to be looking at the right spot at the right moment, that optical burst could have been seen with ordinary binoculars! Shortly after, the star quickly became dimmer and within 10 minutes after the burst it was no longer visible.

The first detection of an optical burst was now a fact. 'We were very lucky', says Akerlof. 'To start with, of course, the weather cooperated. And even though one of the four cameras wasn't working, we managed to catch the optical burst anyway because it was not in the field of view of that camera. Finally, we almost missed the burst completely; it was just a few millimeters away from the edge of the image.' After discussing the results with his colleagues, Akerlof announced the ROTSE discovery through the GCN network around 6 p.m. that evening. Soon congratulations began coming in. 'In all honesty, I must admit that I checked the weather in Livermore and I saw that it had been overcast the whole night', he says.

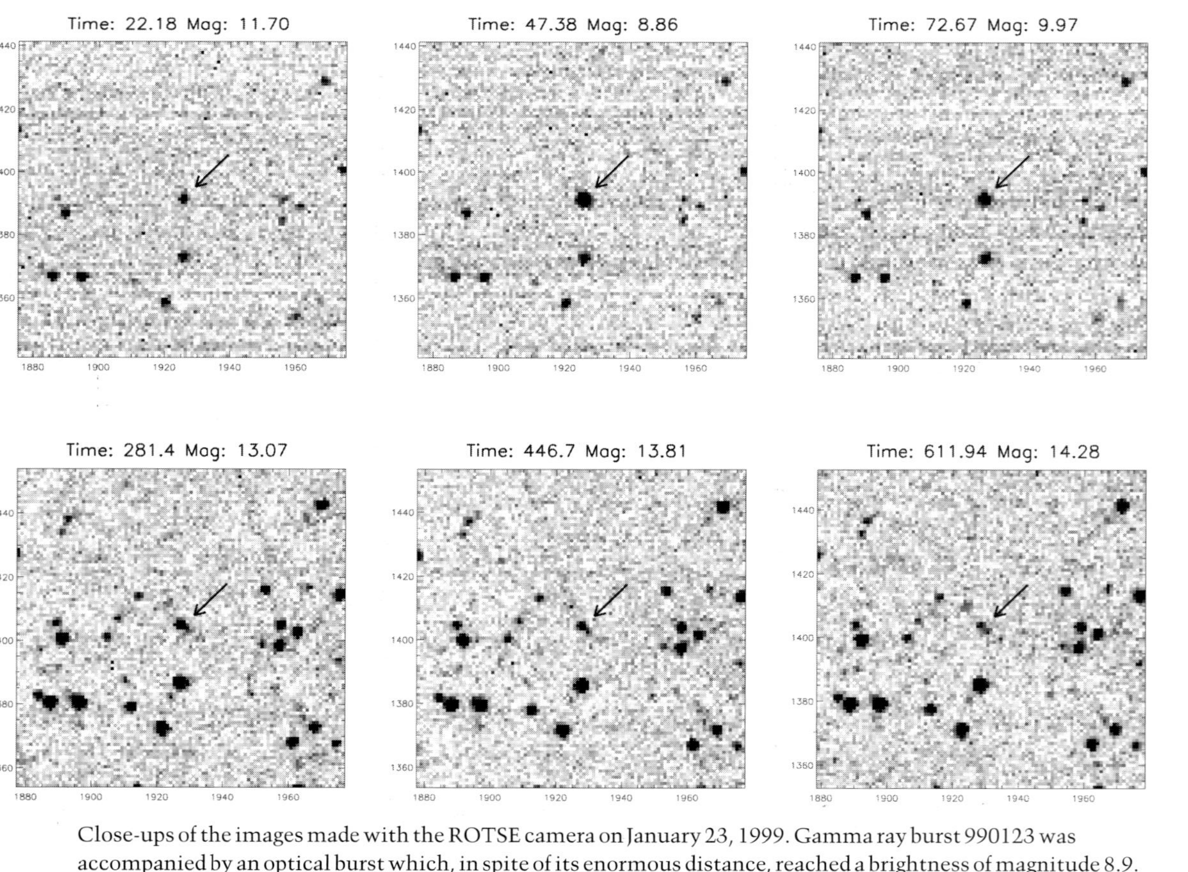

Close-ups of the images made with the ROTSE camera on January 23, 1999. Gamma ray burst 990123 was accompanied by an optical burst which, in spite of its enormous distance, reached a brightness of magnitude 8.9. Above each image can be found the number of seconds that passed since the gamma ray burst.

Hye-Sook Park could have howled when she read the GCN circular. 'If it had been clear, we would have gotten much better observations', she says. 'In the time that ROTSE made three images, LOTIS would have made ten. We would have been able to follow the brightness evolution much more accurately.'

On Monday, January 25th, Akerlof talked with *Nature* about the possibility of publishing, and as they always do, the editors required that nothing be made public about the discovery before publication. Immediately, that requirement went wrong because there are some science journalists who do happen to read the GCN circulars. To avoid the possibility of incomplete or incorrect information appearing in the newspapers, NASA – the biggest money behind Akerlof – decided to issue a press release of its own about the discovery initiated by the NASA-supported BATSE detectors, which led again to many discussions with the editors of *Nature.* Finally, the article was published on April 1, 1999, with Akerlof's name as first author in the list of twenty-one. Hye-Sook Park's name was only mentioned in the small-typed references at the end of the paper. And gracing the front cover of that issue was a photograph of the ROTSE cameras.

Soon afterwards, a second hut was built to house a 45-centimeter telescope right next to the original ROTSE enclosure. This one cannot be aimed as quickly and it has a considerably smaller field of view, but it can detect much fainter objects. According to Akerlof, ROTSE II will make use of the accurate measurements of burst positions provided by NASA's High Energy Transient Explorer 2 (HETE 2). Nor did the instrument remain on the grounds of the Los Alamos laboratory. It has been moved to a much darker spot in the mountains to the west of Los Alamos.

Thanks to the success of the ROTSE experiment, Akerlof received a substantial grant from NASA in the fall of 1999 for a period of 3 years, and plans for ROTSE III have already been made. Akerlof wants to build eight automatic 45-centimeter telescopes erected all over the world so that no burst will be missed.

And Park? She is busy with Super-LOTIS, a 60-centimeter telescope with a very sensitive photometer that can detect objects of magnitude 19 or 20. The first observations have already been made, and when the instrument is fully automated, it will focus mainly on the period between the first 12 minutes and the first few hours after the gamma ray burst, when fast but less-sensitive instruments like ROTSE and LOTIS will have already done all they can do, and the really big telescopes would not yet have had the time to react to a new discovery

At the beginning of 2000, Super-LOTIS still stood in an improvised housing on the parking lot of the Livermore laboratory, covered over with tarpaulin. But now it has been moved to the Kitt Peak observatory in Arizona. 'That optical flash on the 23rd of January 1999 was very exceptional', Hye-Sook Park says thoughtfully. 'The gamma ray burst really was unusually bright but it *is* strange that we haven't observed more optical bursts.' Her eyes suddenly begin to sparkle again.'What we need now are more observations.' Then she starts to laugh and says, 'and what I need is luck'.

14 Fireworks and black holes

'I was watching the boys getting in sticks for the bonfire–and it wants plenty of sticks, Kitty! Only it got so cold, and it snowed so, they had to leave off. Never mind, Kitty; we'll go and see the bonfire tomorrow.'

Stan Woosley has a burn on his left hand. Nothing serious, but it's 6 days since New Year's Eve and it's still visible. In Santa Cruz it is illegal to set off fireworks – California has enough forest fires – but in the first minutes of the year 2000 Woosley couldn't resist it. He doesn't have one good word to say for the unreliable commercial fireworks he fired off that night. 'With my own fireworks such an accident couldn't have happened', he says.

Woosley's passion for anything that explodes and makes a deafening noise dates back to his youth in Texas where he grew up. He used to build rockets and bombs. The bigger the blast, the better. One time he made a mistake and burned down the shed behind his parents' house in Fort Worth but that never happened again. His biggest dream is still to set off the most gigantic fireworks display from a safe place like a pier out at sea.

Anybody who is so obsessed with violent explosions has to be interested in gamma ray bursts. After all, they are the most massive explosions nature can produce. The fact that those cosmic super-bombs are hardly visible from earth does not make them less fascinating for a fireworks fanatic. Woosley, a theoretical physicist at the University of California at Santa Cruz, has been gnashing his teeth about the mystery of gamma ray bursts since 1976 but he now believes he is on the track of a solution. His collapsar model seems to be the best one around at the moment. If Woosley is right, gamma ray bursts are the birth cries of black holes.

When he began to study physics in the early 1960s at Rice University in Houston, gamma ray bursts had not yet been discovered. But astronomers were very familiar with supernova explosions and it was already known that supernovae are the alchemists of the cosmos.

During these powerful explosions the heavy elements of the periodic table are synthesized.

The important role of nuclear fusion in the formation of the elements was convincingly demonstrated in the 1950s by a small group of brilliant astrophysicists under the leadership of Fred Hoyle from Cambridge University. Together with the astronomer couple Margaret and Geoffrey Burbidge, originally from Britain, and Willy Fowler from the California Institute of Technology, Hoyle calculated how all the elements in nature, with the exception of hydrogen and helium, were formed by nuclear reactions within the hot interiors of the stars.

The article that Burbidge, Burbidge, Fowler and Hoyle published in 1957 in the *Review of Modern Physics* had the simple title 'Synthesis of the elements in stars', but it went down in history as B^2FH, the initials of the authors. It is regarded as one of the most important publications in the history of astronomy. But even though Hoyle, Fowler and the Burbidges laid down the basis for the theory of cosmic nucleosynthesis, there were many details that were not filled in during the 1960s.

Very early in his student days, Stan Woosley set his mind on the cosmos. 'President Kennedy visited our university', he said, 'and told of his plans for the Apollo program. There was going to be a big space flight center in Houston so I didn't want to leave yet.' With Don Clayton and David Arnett, he did research on explosive nucleosynthesis. Clayton was one of Fowler's students and he continued Fowler's pioneering work on the origin of the elements.

Woosley can still go into raptures when he thinks back to that time. As a child of 12 years of age, he already knew from memory the periodic table that every student remembers seeing on the chart hanging in the chemistry room at school, and here he was, trying to unravel the history of how those elements came to be. 'It took a long time, but we got it!', he says.'For every isotope[1] from every chemical element we

[1] Each element is distinguished by the number of positively charged protons in the nucleus of the atom. However,the number of uncharged neutrons is not always the same. Most elements have various isotopes: atoms with the same number of protons but with a different number of neutrons in the nucleus and with a different atomic mass.

(*Left*) *Stan Woosley* developed the collapsar model in which a gamma ray burst is the result of a catastrophic collapse of a super-massive, rapidly rotating star into a black hole.

(*Right*) *Andrew McFadyen* developed computer software in which the catastrophic collapse of massive stars can be simulated.

were able to get behind the history of its origin. Every atom has its own interesting story to tell.'

For many heavy elements and rare isotopes that story can be unusually complicated. The nuclei of atoms can fuse together, split in two, be bombarded by high energy neutrinos; they can be radioactive and can lose one or more nuclear particles, or they can even capture a neutron and in this way become a heavier isotope. And many of these processes can only take place during supernova explosions. Woosley was on the track of understanding the makeup and machinations of the most beautiful fireworks in the cosmos.

The night sky with its twinkling stars may seem serene and constant, but for Woosley it is everything but. It was in the 1960s that explosive quasars and high-energy pulsars were discovered and the mechanism of catastrophic supernovae was unraveled. 'The universe is a mass of violent action', says Woosley. The enormous amount of energy that can come free in such a short time still amazes him. 'When a massive star explodes, it releases the same amount of energy in one second that the sun sends out in billions of years. And if you add to that the fact that the sun produces in a hundred thousandth of a second as much as all the nuclear weapons on earth, it truly boggles the mind.'

Woosley's thesis had to do with the forming of all the elements from magnesium to nickel. It is still regularly referred to, which is something he is proud of. After graduating in 1971 he worked for 3 years in Pasadena with Willy Fowler who, in a sense, was his academic 'grandfather'. In 1975 Woosley left for Santa Cruz, but a picture of Fowler, who won the Nobel prize for physics in 1982, still hangs on the wall of Woosley's office. Fowler died in 1995.

For all the research into the numerous nuclear reactions that are responsible for the formation of heavy elements in the supernova phenomenon, it is still far from being fully explained. Theoreticians want to know exactly how the collapse of the core of a star comes about, how the pressure and temperature are distributed, at what moment outward shock waves are produced, and how the flood wave of neutrinos from the very innermost part of the dying star succeeds in turning the implosion to an explosion. In the course of the 1970s and 1980s, Woosley worked closely on this with Tom Weaver from the Lawrence Livermore National Laboratory. Late at night when the supercomputers were able to rest from their labor-intensive calculations for nuclear weapons, Woosley and Weaver put them to work calculating the last evolutionary stages of massive stars.

Shortly after starting work at the University of California, Woosley got interested for the first time in gamma ray bursts, the discovery of which had been announced in 1973. 'In 1976 I published my first gamma ray burst model', he says. 'According to that model, the burst was produced by thermonuclear explosions on neutron stars upon which the matter from an accompanying star had landed.' Woosley says the model was very popular for about 10 years, but finally it could no longer be reconciled with the observations, especially when it became clear that gamma ray bursts occur at much greater distances than a few hundred light years away.

At the beginning of the 1980s it was clear that the gamma ray burst mystery would only be solved when the explosions were also studied at other wavelengths. And so, during a workshop in Santa Cruz in July 1983, the idea evolved for NASA's High Energy Transient Explorer (HETE), a name that Woosley dreamed up. With the help of ultraviolet cameras and X-ray telescopes it would be possible for HETE to fix the positions of the gamma ray bursts accurately and, hopefully, to determine their distance.

Thirteen years later, in November 1996, HETE was launched and shortly thereafter got lost (see Chapter 4). 'It was horrible', says Woosley, who had never been so closely associated with a space research program before. 'It's as if you've lost your child. I have renewed respect for those people who earn their living working in space research.' A few months later BeppoSAX discovered the first X-ray afterglows and Jan van Paradijs' team in Amsterdam discovered the first optical afterglow from a gamma ray burst.

Woosley has mixed feelings about BepoSAX. Naturally, thanks to the Italian–Dutch satellite, the research really took off and the interest in gamma ray bursts increased enormously. But he had to come to terms with the fact that this impressive breakthrough came from Europe rather than the US, and that the BeppoSAX team won the Bruno Rossi prize in 1998. Nevertheless, when BeppoSAX got into serious financial difficulties at the beginning of January 2000, Woosley was very active in the campaign to 'Save Sax', which led to keeping the satellite alive and operational for another year.

Woosley is also closely involved with HETE 2, the successor to HETE, but above all, he is a theoretician whose mind is constantly busy building models. After various versions of the thermonuclear explosion model, he also worked on a model in which gamma ray bursts come from the impact of asteroids on neutron stars. That model was also thrown into the trash basket, but that's par for the course. You can recognize a good theoretician when his wastebasket is fuller than his desk, because that means he has explored all the conceivable possibilities. Nevertheless, Woosley hopes that his 'collapsar' model is destined to live longer.

The word 'collapsar' combines 'collapsing' and 'star'. A collapsar is a super-massive, rapidly rotating star that collapses at the end of its life to become an all-devouring black hole. Nevertheless, in the forming of that black hole, a tremendous amount of energy comes free. This energy is radiated outward in two narrow beams as gamma rays, and when the earth happens to be in line with one of these beams, we see a gamma ray burst flare up in the sky.

The first ideas about collapsars came during the 1980s. Together with his colleague Peter Bodenheimer, Woosley made theoretical model calculations on supernovae. The great mystery about supernovae has always been that the star blows apart through the force of gravity, instead of imploding upon itself, as you would expect. This could be explained in part by the shock wave that is created when the collapsing mantle of the star is spattered into smithereens on the incompressible surface of the newly formed neutron star. But calculations have shown that this shock wave, in most cases, is not strong enough to have such an effect. The enormous flood of neutrinos that is produced when a neutron star is formed seems to play a crucial role in the making of a supernova explosion.

Nevertheless, the model builders, according to Woosley, often have trouble making a model that would lead to an explosion. Depending on the properties of the star, such as its mass and rotational speed, the calculations often end in a catastrophic collapse without the mantle of the star being blown out into space. 'Our goal was to improve the fit of the models so that the explosion would actually take place', says Woosley, 'but after a while, I began to ask myself if it were not possible that in some cases, the calculations were right. Imagine that a black hole forms; what exactly happens then?'

One thing was sure: if no neutron star is formed in the interior of the collapsing star, then there is also no outwardly directed shock wave. Moreover, there is no flood of neutrinos, which means that supernova explosions fail to appear. It looked as if extremely massive stars – more than forty times as heavy as our sun – could indeed end their lives in such a 'failed supernova'. Only when a dying star turns rapidly on its axis could an explosion still take place.

In the beginning Woosley did not think at all about gamma ray bursts when he first put his 'failed supernova' model down on paper. That only happened in the early 1990s when Bohdan Paczyński from Princeton University spoke about his own ideas at a symposium. Paczyński, who was absolutely convinced that gamma ray bursts originated at great distances in the universe, thought that they had to do with the merging of neutron stars.

When both components of a binary star system experience a supernova explosion, a binary neutron star is left over. Our own Milky Way galaxy has several such binary neutron stars. The two compact objects orbit each other rapidly and spiral slowly but surely inward: their mutual distance becomes smaller and smaller and the orbital speed increases. That effect, which is predicted by Einstein's theory of general relativity (the binary star system loses energy when it sends out so-called gravitational waves) was observed for the first time in 1974 by Paczyński's colleagues Russell Hulse and Joe Taylor, who won the Nobel physics prize in 1993 for this work.

If binary neutron stars do indeed spiral towards one another by sending out gravity waves, then they must ultimately merge with each other. For the famous binary pulsar PSR 1913+16 that Hulse and Taylor had studied, this will occur in about a hundred million years. If such mergers occurred in our own Milky Way galaxy they would also occur in other galaxies, according to Paczyński. And because the number of galaxies in the universe is so incredibly great, such happenings probably occur at the rate of one every day.

The merging of two neutron stars is a most violent phenomenon. The two small, compact stars are torn apart by gravity and disappear in a black hole. But that doesn't happen easily. Because the neutron stars are spinning around each other at high speed, the stellar gas first piles up into a rapidly rotating accretion disc before disappearing for good into the black hole. According to Paczyński the high-energy phenomena in that accretion disc could lead, in one way or another, to a gamma ray burst.

Woosley realized that in his 'failed supernova' model, something similar is happening. If the star turns fast enough on its axis, then it

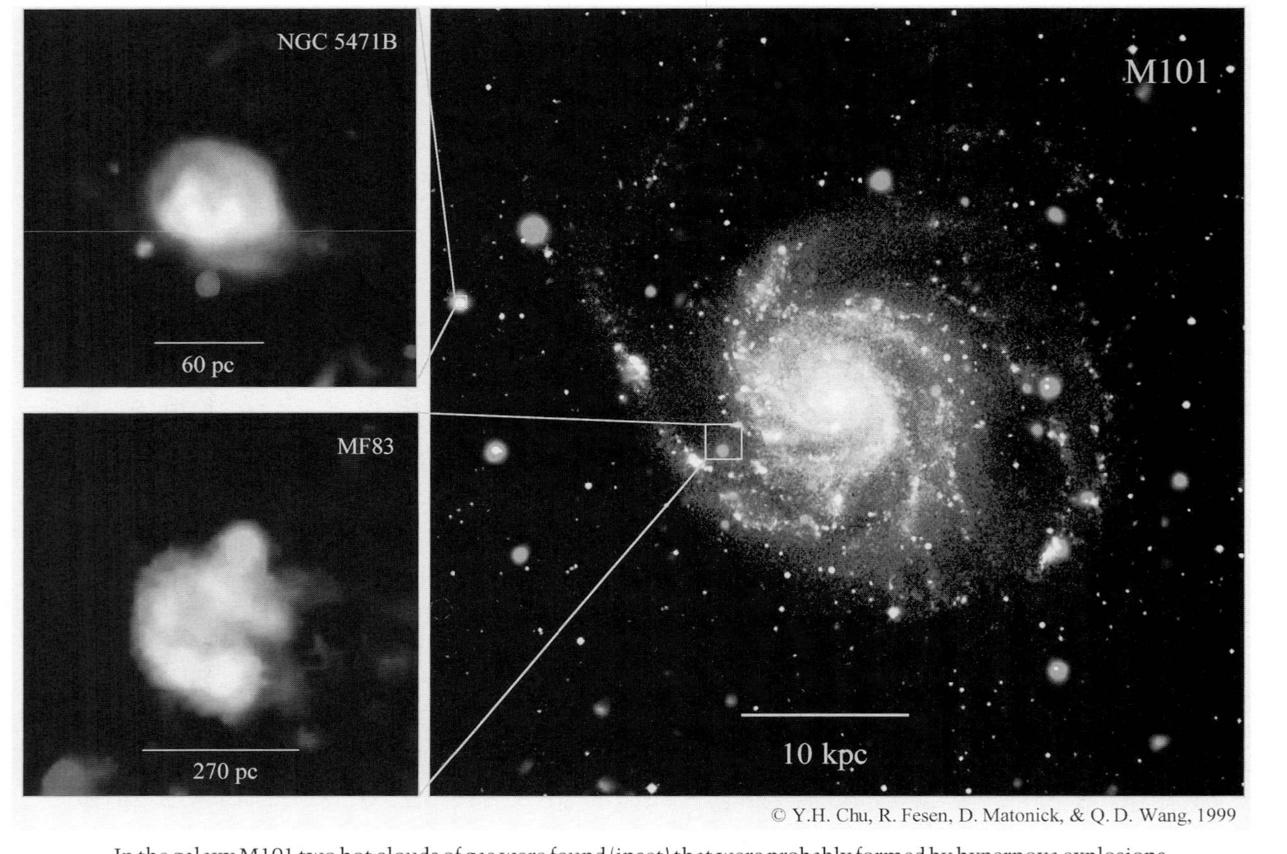

In the galaxy M101 two hot clouds of gas were found (inset) that were probably formed by hypernova explosions.

would also develop an accretion disc. And if along the rotational axis of such a disc matter can be blown out at extremely high speed, it is possible that a gamma ray burst could be produced. In 1993, he described his new model in *The Astrophysical Journal*.

'I named my model the failed-supernova model', he says. 'That was a terrible mistake. Since gamma ray bursts are the most energetic explosions that have ever been observed in the universe, the word, "failed" is out of place. "Hypernova" would have been a much better description, but I had already used this term in 1982 to accommodate a particular kind of supernova.' However, this was of no concern to Paczyński, who did use 'hypernova' because he didn't really care in what way the explosion actually came to be. 'Though I stopped resisting the word "hypernova",' says Woosley, 'I still prefer the term "collapsar" for my model, because it tells you something about the underlying mechanism of gamma ray bursts.'

In 1993, the calculations for collapsars were still rather primitive, and according to Woosley based primarily on simple algebra. There was no talk of detailed computer simulations. But things changed quickly when Andrew MacFadyen came to Santa Cruz. MacFadyen let supernovae explode on his computer screen and conjured up, with the same ease, a collapsar.

Andy MacFadyen grew up in New York and ever since he was a little boy he was mesmerized by the fascinating world of the universe. He devoured books about the big bang, higher dimensions and the endlessness of the cosmos. He was also a great fan of Carl Sagan's TV series, *Cosmos*. 'Since I was 7 years old I knew I wanted to be a physicist or an astronomer', he says. An inspiring physics teacher did the rest. After studying astrophysics at Columbia University in New York and Harvard in Cambridge, Massachusetts, MacFadyen moved to the west coast of the United States in 1992 where he went to the University of California at Santa Cruz to work on supernovae.

During the ensuing years he developed a computer program with which he could imitate the evolutionary stages of massive stars in the greatest detail – a three-dimensional simulation program that can be played around with at will. 'Astronomy is usually a rather passive science', says MacFadyen. 'You can only look at what is actually happening. But with a computer program you can experiment to your heart's content. You can hold nature in the palm of your hand and you call the shots. What happens if you blow up a star; if you let it rotate incredibly fast? Or if you make a star out of green cheese?'

But holding nature in your hand sounds more profound than it is. Nature, of course, does not take her lessons from a theoretical astrophysicist on earth. MacFadyen calls his simulations simply 'stars' but the best you can hope for is that they are a fair approximation of reality. 'As a theoretician you must always be aware of the fact that you are working with models', he says, 'no matter how believable they seem.' And it has to be borne in mind that you always work with qualitative calculations that may only be able to bring general principles to light, just as meteorological models of the earth's atmosphere correctly predict large-scale flow patterns but not the exact moment and place it will start raining.

Incidentally, there is more agreement between MacFadyen's simulations and the models of meteorologists than you would expect. Hydrodynamic processes play an important role not only in the atmosphere of the earth but in the interior of a giant star as well. There are complex calculations that must be very accurately and efficiently figured out. 'But', says MacFadyen, 'I love waves; I love oceans; I love flow patterns; that's why I like this work. It's an amazingly rich field for research: star formation, colliding galaxies, cosmology – anywhere you look you need hydrodynamics.'

MacFadyen works very closely with Woosley on collapsars – those super-massive stars that rotate rapidly, and at the end of their life collapse into a black hole. In a small, dark corner of his office he spends his days, for the most part, behind his computer monitor. His long black hair is tied back in a pony-tail and he listens via his walkman to old 1950s rock-and-roll music. On the screen there are beautifully colored patterns that make you think of the fluttering wings of butterflies and then again, of the undulating unfolding of a flower. The psychedelic patterns look like they come from the flower power culture of the 1960s.

In the simulations everything is taken into account: the composition of the star, the temperature, the pressure and viscosity of the stellar gas, the nuclear reactions that take place internally and, of course, the all-controlling force of gravity. In small time increments it continually calculates how the star evolves under the influence of all these forces. The result of one calculation forms the input for the next. By the time the interior of the star begins to collapse, the time increments must not be larger than a few microseconds, otherwise the simulations are no longer accurate enough. Because the many calculations for each time increment also takes the computer a couple of microseconds to process, the formation of a black hole develops on the screen at about the same rate as it does in actuality. 'That's what I find so great about models that evolve in time', says MacFadyen, 'they seem so beautifully realistic.'

For someone who works day in and day out with these models, that may be the case. But for the layman the moving red, yellow and green patterns don't seem to have much to do with what is happening in the deep interiors of dying giant stars. The colors are used to indicate the differences in pressure, temperature or gas density. The monitor shows only the innermost 1% of the star, where the evolution occurs most quickly. 'On this scale the star itself would be bigger than this room', says MacFadyen. 'This is zoomed in enormously.'

Woosley is just as enthusiastic about the simulations his graduate student is making as MacFadyen is himself. Of course it is 'just' a simulation. The real thing may look completely different, but broadly speaking, Woosley has a tremendous amount of faith in the models, and not without justification. The supernova explosion that flared up on February 23rd, 1987, in the Large Magellanic Cloud, which was the most intensively studied supernova in the history of astronomy, behaved exactly as Woosley's (older) theoretical predictions said it should.

The collapsar drama begins with a rapidly rotating star that is at least forty or fifty times as massive as the sun. Such a giant star burns up its nuclear fuel very quickly and has a lifetime of only about 10 million years. In the last stages of its life the nuclear reactions in the interior of the star follow one another in rapid succession (see Chapter 11) and finally a core of iron atoms is formed in the center of the star that is twice as massive as the sun.

The destiny of the star is now sealed. No spontaneous nuclear reactions can occur any more in the iron core and the internal radiation pressure that acted against the force of gravity drops away. The star's core collapses into a 30-kilometer-large neutron star but it is not able to reverse the gravitational collapse of the star. Every second a few hundred thousand trillion trillion tons of stellar gas plummet inward and nothing can stop it now. Moreover, the neutron star is so massive that it collapses even further under its own weight into a black hole with a diameter of about 10 kilometers. As a result, there is no shock wave sent out nor is there a flooding of high-energy neutrinos.

You would expect that the star would be swallowed up by the central black hole in a few seconds, but MacFadyen's computer simulations show that the star does not disappear from the stage completely. The rapid rotation of the original giant star plays a decisive role. With the collapse of the stellar core the rotational speed increases enormously to many hundreds of rotations per second. The gas in the equatorial plane of the star cannot disappear directly into the black hole because it is rotating too fast. This creates a thick equatorial accretion disc with a diameter of about 100 kilometers. The temperature in the disc reaches more than 10 billion degrees; the density is about 10 thousand tons per cubic centimeter and the rotational speed can be more than 1000 revolutions per second.

Along the rotational axis of the star the gas can much more easily disappear into the black hole. The force of gravity sucks the gas into the black hole with enormous speed, resulting in the formation of funnel-like areas above the north and south pole, in which the gas density is much lower than in the immediate surroundings, as if the region around the rotational axis is being evacuated by the black hole.

Meanwhile, a huge amount of energy is generated in the accretion disc. The density of the stellar gas is unbelievably high and while the gas is spiraling into the black hole a horrific amount of friction develops. The accretion disc begins spontaneously to radiate elementary particles and anti-particles; in fact, energy is turned into matter according to Einstein's formula $E=mc^2$, and those particles, moving almost with the speed of light, can only escape along the rotation axis because the gas density is much lower there. This creates two oppositely directed jets that come from deep within the star and they propel their way to the outside with tremendous energy.

Woosley admits that they cannot be certain about all the details. It is possible that strong magnetic fields also play an important role. But magnetohydrodynamics (MHD) is such an incredibly involved subject that truly accurate models cannot yet be made. Many astronomers say that the acronym might better stand for 'magic hydrodynamics'. 'However, it seems just about impossible *not* to produce jets', says Woosley. 'No matter how the energy is created, it gets sent out in two beams.'

The jets find their way to the outside very quickly and blow apart part of the star. This happens only when the star, at the end of its life, has already blown its outermost hydrogen-rich gas layers into space, otherwise the two jets would come to a halt in the thick mantle of the star. But when everything goes right – which means that there is a large mass to start with, a high rotational speed and a considerable loss of mass in the last evolutionary stages – then the collapsar eventually spews matter into the universe in two directions with almost the speed of light. Shock waves in the beams and in the surrounding interstellar matter produce high-energy radiation, and anyone who is in the path of one of the two jets will see a powerful gamma ray burst flare up.

The hypnotic film on Andy MaFadyen's computer monitor comes to a stop. The star is dead; the gamma ray burst goes forth to the furthest reaches of the universe as if an obituary is sent around with the speed of light. MacFadyen leans back in his chair, satisfied. Woosley still seems to be lost in thoughts of the biggest fireworks display that the cosmos has to offer. Outside, it is twilight and most of the colleagues of the two supernova gurus at Santa Cruz have gone home. Apart from the hum from MacFadyen's computer, it is still; as still as the dark burial ground of the dead star.

Do the computer simulations form a reliable representation of reality? Woosley thinks so. 'You can ask yourself if this model offers a good explanation for *all* gamma ray bursts', he says, 'but I am convinced that the longer bursts, of more than a few seconds, are caused in this way. If this model doesn't fit, then there must be something even crazier going on.' In the coming years Woosley and MacFadyen, who received his PhD in the beginning of 2000, hope to extend and refine their calculations. The ultimate goal is to design models that will give predictions in such detail that they can be immediately compared to the observations of actual gamma ray bursts and supernovae.

However, there is already observational support. If the collapsar model is correct, gamma ray bursts are expected to occur primarily in star-forming regions. Massive stars live for only a short time and one could say that their grave is right next to their cradle. In 10 million years they cannot travel distances of thousands of light years and when they collapse into a black hole they will still find themselves in the gas- and dust-rich star-forming regions where they were so recently born. Observations by the Hubble Space Telescope of the host galaxies of gamma ray bursts indeed show that the super-explosions take place in active star-forming regions.

Moreover, both BeppoSAX and XMM-Newton, a large European X-ray observatory in space, have found strong evidence for the existence of large amounts of iron in the immediate surroundings of a few gamma ray bursts. Supernova-like explosions are the only known sources of iron in the universe, so the observations seem to indicate that the gamma ray bursts are associated with exploding stars.

In contrast, the association with star-forming regions is bad news for the theory that gamma ray bursts are produced by the merging of neutron stars. Although a binary neutron star may form quite soon after the birth of the original binary star, it takes many tens, or even hundreds, of millions of years before the two neutron stars spiral towards each other and merge into a black hole. They would then have left the dusty star-forming region where they were born long ago. Not only does each star describe its own orbit through the galaxy in which it finds itself, but asymmetrical supernova explosions can also 'launch' a newly born neutron star in a particular direction at very high speed. If gamma ray bursts were created by the merging of neutron stars, you would expect to find them everywhere, even in the space between the galaxies.

Nevertheless, Woosley thinks that some gamma ray bursts are due to the merging of neutron stars. After all, binary neutron stars do exist, and radio observations have convincingly shown that they spiral in towards each other. There is no doubt that neutron stars must bang into each other with great regularity everywhere in the universe. That is another way in which black holes are produced and the birth cries of these black holes must be observable in the form of gamma ray bursts.

Perhaps the 'long' gamma ray bursts (longer than about 2 seconds) are the result of the catastrophic collapse of massive giant stars – collapsars – and the short flashes are produced by colliding neutron stars. This would be a lovely explanation for the fact that the thousands of gamma ray bursts that have been observed up until now clearly fall into two groups that not only differ in the average duration but also in the energy of the gamma rays that are sent out.

NASA's High Energy Transient Explorer 2, the successor to the original HETE satellite, is expected to shed some light on the subject. Up until now no afterglows of short gamma ray bursts have been observed, and so naturally no position in the sky could be fixed. But HETE 2, which was launched on 9 October 2000, should change all this. Through the combinations of gamma ray detectors and X-ray telescopes the sky position of the bursts can be determined within a couple of seconds and sent through to observatories and robotic cameras on earth. If there are different ways in which black holes are formed and if there are different sorts of cosmic fireworks, then HETE 2 will shortly find out what is happening.

On the campus of the University of California at Santa Cruz situated high up in the hills, night has fallen. Down below, about a couple of kilometers away, the dim lights of the city are glowing; above, at a distance of many tens or hundreds of light years away, the little lights of the stars are twinkling. Suddenly a silent arrow of fire shoots across the sky, blue-white in color and brighter than the stars in the heavens or the little lights along the coast. Not a festive fireworks rocket, but a meteor, which at a height of 80 kilometers cuts a glowing trail of light through the atmosphere.

A street light can no longer be seen at a distance of 80 kilometers and the bright meteor would be totally invisible at tens or hundreds of light years away. And as far as the stars in the sky are concerned, nothing would remain to be seen if they were billions of light years away. But out there, in the farthest, darkest reaches of the universe, a fight to the death takes place every now and then, releasing so much energy that we, way down here on earth, could bear witness to an almost eye-blinding burst, if only our eyes had the least bit of sensitivity to gamma rays.

The fireworks that celebrate the birth of black holes overshadow every other light show the heavens have to offer. And as a matter of fact, they have made Stan Woosley's dream of the perfect exhibit come true: a gigantic fireworks display at a safe distance that behaves exactly as the fireworks master has calculated it should. The only downside to enjoying the show is that you would have to look through artificial eyes in orbit around the earth.

15 A flashing future

'I know something interesting is sure to happen', she said to herself.

Sunday June 4, 2000, was a black day in the life of Neil Gehrels. Above the Pacific Ocean, 4000 kilometers southeast of Hawaii, one could have seen a spectacular cosmic light show that morning. An enormous fireball shot across the sky as if a giant meteor had penetrated the atmosphere of the earth. At a height of many, many kilometers the projectile broke into pieces due to the inconceivable friction that was generated as it entered the atmosphere, and the fireball shattered into a myriad of smaller meteors. A little later the scorching fragments plunged into the water and sank down at once to the ocean floor.

But these were not real pieces of cosmic rock that fell from such great heights to lie deep beneath the surface of the sea. They were the partially molten remains of the Compton Gamma Ray Observatory (CGRO). Some of the pieces of metal were no bigger than gravel; other fragments weighed a few hundred kilograms. Compton, an observatory for cosmic gamma rays, was given a sailors' burial at sea. And America's NASA had prematurely condemned the observatory to death.

Gehrels had been the project scientist of CGRO since its launching on April 5th 1991. For more than 9 years the satellite had done valuable research on solar flares, supernovae, exploding galaxies, quasars, and of course, gamma ray bursts. The scientific instruments, designed for a lifetime of 5 years, were still working perfectly; gamma ray detectors have no moving parts, which means that they don't get worn out. There are those who believe the Compton could have continued to function for another 10 years.

Even so, the colossal satellite, with malice aforethought, was pulled down in June 2000. Protests from scientists were to no avail. NASA had so decreed, and even though, 5 months earlier, Gehrels secretly believed that NASA would change its decision, by the middle of May he gave up all hope. On Sunday June 4th the curtain fell on the Compton observatory.

Why bring down a satellite that is still working perfectly? According to NASA it was always a question of safety. With a weight of almost 17 tons, Compton was the heaviest scientific satellite that had ever been put into orbit around the earth. If it were left to its fate, it would, like all other satellites, eventually enter the earth's atmosphere. This happens because the satellite gradually loses some orbital energy as a result of the friction with the atoms in the very highest, thinnest layers of air. But while 'ordinary' satellites burn up completely when they enter the earth's atmosphere, CGRO was so large and heavy that fragments could smash into the earth, possibly in an inhabited area.

It had always been NASA's plan to bring the gamma ray satellite down before its supply of fuel ran out. That fuel was intended for small steering rockets that could 'kick' the satellite into a higher orbit if necessary and that were also used for altitude control. With help from the steering rockets and gyroscopes the satellite could be directed to a safe place so that the chance of hitting humans was just about zero.

But in December of 1999 one of the gyroscopes on board CGRO gave out. With the help of the other two remaining ones, a safe descent could still be made, but if another gyroscope failed that would not be possible any more, according to NASA. The satellite had to be brought down before it was too late.

The decision came as a shock to many scientists who were closely associated with the Compton project. 'I never realized that the loss of one gyroscope could have such dire effects', says Jerry Fishman from NASA's Marshall Space Flight Center, principal investigator for the BATSE detectors on board the Compton satellite.

Technicians from the Goddard Space Flight Center in Greenbelt did not sit idly by. In the beginning of 2000 they developed, in a short time, a method of steering Compton reasonably well without any gyroscopes at all. Now there was no need to take the heavy satellite out of orbit, it seemed. But NASA was not about to take any risks. At a press conference on March 24th, Ed Weiler, NASA's director of science, explained why Compton must be brought down. The Goddard method (without gyroscopes) would increase by a factor of seven the chance of someone on earth getting struck as compared with the NASA 'safe' method of using two gyros. Taking that greater risk would be irresponsible, according to Weiler.

But Jim Ryan from the University of New Hampshire, at Durham, one of the principal investigators of COMPTEL (another gamma ray detector on board the satellite), dismissed this conclusion out of hand. The numbers would probably be right, but in both cases the risk was absolutely negligible, according to Ryan. With two gyroscopes working, the NASA risk analysts declared that there was an infinitesimal chance of 1 in 29 million that somebody on earth would be hit by a fragment from the satellite. With the Goddard method the chance was 1 in 4 million. 'But that would mean that for 11,000 years, every day a satellite like Compton would have to come down before there would be one victim', says Ryan. 'Many other NASA activities, such as the launching of the space shuttle, are much more dangerous.' Even coming to the press conference on March 24th carried more risk, in the eyes of Ryan. The chance that a journalist would have a fatal car accident in Washington was greater than the chance that Compton's gyroscopeless trip back to earth would cost someone their life.

Ryan mobilized a number of colleagues and started a protest action with the hope that NASA would change its mind. They not only wanted to have their arguments heard by NASA headquarters in Washington but also wanted to reach Congress through a letter-writing campaign. But it didn't work. In 1999, NASA had a bad image because of the recent loss of two Mars probes, and no matter what it cost, they wanted to avoid giving the impression that they were not serious about human safety.

Gehrels had given up a while back. 'No matter how unhappy I am with the decision, NASA has done its work conscientiously', he said in the middle of May. Maybe the decision would have been different if Compton had recently been launched, thought Gehrels. But the satellite had already functioned twice as long as originally planned and most of the scientific goals were achieved. Now one had to weigh between science and safety.

Neil Gehrels' name is actually Cornelis, after his father's oldest brother, who died in a German concentration camp in 1945. Both of his parents come from the Netherlands. Neil's father, Tom Gehrels, is a well-known astronomer at the University of Arizona. Not that astronomy was fed to their offspring with their baby food; Gehrels senior wanted his children to learn to be independent and make their own choices. In high school Neil tried to avoid physics and astronomy as much as possible; he wanted to be a composer. But the music suited him better as a hobby than a profession, and so at the University of Arizona he indeed ended up studying physics.

It was while he was a student that the discovery of gamma ray bursts became known around the world, but it didn't reach Gehrels. In 1973, when Ray Klebesadel presented his Vela observations, Neil and his younger brother George were hitch-hiking around the world. In 1975, when Jerry Fishman carried out his first balloon flights, the brothers were traveling by bike through Mexico and Central America, a trip of about 4500 kilometers. The world had so much to offer; the cosmos could wait.

After doing research for his thesis at the California Institute of Technology from 1976 to 1981, Gehrels left for the Goddard Space Flight Center where he was to work on GERSE (Gamma-ray Energetic Radiation Spectroscopy Experiment), the gamma ray spectroscope that would be placed aboard the Compton Gamma Ray Observatory. But one year later, GERSE fell victim to budget cuts. 'Ultimately that turned out to be good for me', says Gehrels. 'I was forced to learn more about gamma ray spectroscopy which had a more significant role in the balloon instruments that replaced the GERSE program.'

Via balloon experiments and the development of instruments for the Nuclear Astrophysics Explorer (a NASA satellite that was never realized) Gehrels got involved with the preparations for the European Integral (International Gamma Ray Astrophysics Laboratory) project, a large gamma ray satellite that is scheduled to be put in orbit around the earth in 2002 or 2003. In the spring of 1991, at the age of 38, he was appointed project scientist of the Compton Gamma Ray Observatory just around the time of its launch.

'I was quite young for such a job', says Gehrels. 'I took it over from Donald Kniffen who had been working on CGRO for about 10 years and wanted to move on to university teaching. The principal investigators from the various CGRO instruments, all of whom had their say in choosing a successor, were very friendly.'

Compton dominated Gehrels' life for almost 10 years, and BATSE was one of his favorite instruments. 'BATSE was absolutely unique,' he says. 'No other instrument was able to constantly monitor the sky in such a high energy range.' The BATSE discovery that gamma ray bursts are distributed equally over the whole sky is, according to Gehrels, one of the most surprising results of the whole CGRO mission.

Naturally, Gehrels was completely aware that an actual breakthrough would only come to pass when gamma ray bursts were also observed at other wavelengths. But chasing optical bursts and afterglows, and searching for afterglows at radio and X-ray wavelengths, was not very popular in the mid-1990s. Only a handful of astronomers busied themselves with that.

In hindsight, the first identification of a gamma ray burst could well have been made by American astronomers rather than have it pass by right under their noses. If the bright optical flash from GRB 990123 had happened a few years before, it would have been discovered with the GROCSE camera in Livermore, and had the German X-ray astronomers reacted more enthusiastically to the requests of Gehrels and Kevin Hurley, the first X-ray afterglow would have been found with the Rosat satellite.

The big problem with BATSE was that the sky positions were not accurate enough and the problem with the Compton satellite was that it had no X-ray detector or optical telescope aboard. HETE would be an improvement, but because it would only monitor a small part of the sky, it would be unable to detect more than some tens of gamma ray bursts a year. Moreover, the X-ray cameras on the small inexpensive HETE satellite are not very sensitive, so that astronomers would still have to make use of telescopes on earth when they wanted to find optical afterglows.

So it was that, around 1994, Gehrels and his colleagues developed their first ideas for a successor to BATSE and HETE. It would be a satellite that would detect at least 300 gamma ray bursts each year. In addition, it would have a battery of different kinds of telescopes that could be accurately pointed within a minute; there would be an X-ray telescope, an ultraviolet telescope and an optical telescope. Such a specialized satellite would fit in well with NASA's Explorer program.

NASA has two kinds of Explorer satellites: the Small Explorers (SMEX) and the Medium-class Explorers (MIDEX). In both cases they are relatively small, inexpensive projects that are directed toward one specific area of research. SMEX projects must not cost more than 80 million dollars and MIDEX projects have a cost cap of about 160 million dollars. In principle, NASA plans to launch one SMEX and one MIDEX satellite each year. Researchers could submit proposals and a selection committee would choose the winning projects.

In the fall of 1997 there were to be two SMEX projects selected and all proposals had to be in by the end of 1996. 'Naturally the competition was fierce,' says Gehrels. 'From the 50 proposals sent in, four or five had to do with research on afterglows from gamma ray bursts. The whole community realized that there was a need for a specialized gamma ray burst satellite.' Gehrels's own proposal was called BASIS (Burst Arc Second Imaging and Spectrometry). The satellite would give the position of bursts with an accuracy of an arcsecond and at the same time do spectroscopic research.

Gehrels made sure that he would have the cooperation of a great number of colleagues, including Scott Barthelmy, Tom Cline, Ed Fenimore, Jerry Fishman, Kevin Hurley, Chryssa Kouveliotou, Jan van Paradijs, Brad Schaefer and Stan Woosley. An imposing list of names, but still the competition was stiff. Charles Haley from Columbia University in New York had also submitted a proposal for a gamma ray burst satellite and worked closely with such people as Shri Kulkarni and Mark Metzger from Caltech and Bohdan Paczyński from Princeton. Haley's satellite was called BOLT (Broadband Observatory for the Localization of Transients).

In the end, neither one was chosen. For the 1997 SMEX project, NASA chose the High Energy Solar Spectroscopic Imager (HESSI), a satellite for solar research, and the Galactic Evolution Explorer (GALEX) which, with the help of an ultraviolet telescope, will be used to study the evolution of galaxies. At the moment it looked like BOLT had the edge on BASIS. The Columbia satellite was officially selected as the backup in case HESSI or GALEX would run into too many delays. BOLT was even being considered as a possible exception to the rule by being chosen as a third project.

However, by the end of 1997 the whole field of gamma ray bursts had changed dramatically. BeppoSAX had discovered the first afterglows; the redshift and the distance for some gamma ray bursts had been determined and the satellite designers knew, as they had never known before, exactly what they were looking for. Both Gehrels's and Haley's groups began to review their projects. Now that they had the opportunity to submit proposals for a new MIDEX project, they had more financial maneuverability.

Gehrels still has to laugh when he thinks back to the speech NASA's administrator, Daniel Goldin, gave at the winter meeting of the American Astronomical Society in Washington in January 1998. 'During his talk Goldin spent a few minutes on gamma ray bursts', he says, 'and with great conviction he declared that NASA would solve the riddle once and for all, probably with a MIDEX satellite. However, according to NASA headquarters, he shouldn't have mentioned it because the MIDEX projects were supposed to be chosen by an independent selection committee. But of course we thought it was wonderful. Later on, I often quoted Goldin's statement in my own presentations.'

The new satellite project from the Goddard team was called Swift, a name that was suggested by Gehrels's colleague Nick White, who played a key role in making the whole project come about. 'No, it isn't an acronym', says Gehrels. 'The name has to do with the fact that the satellite can react so fast.' Again, they were in competition with the Columbia–Caltech team who came up with a new version of BOLT. But there were many other proposals for gamma ray burst satellites, among which was one from the Naval Research Laboratory, working together with the group in Huntsville and the BeppoSAX team in Rome. Incidentally, Gehrels also worked with the Italian astronomers. In exchange for observing time on the new satellite, the Brera observatory near Milan offered a telescope mirror at no cost.

The normal procedure for making a MIDEX selection from the many proposals submitted required that five would be chosen for what is called the phase A study. From these five, there would be two definitely selected on the basis of a much more detailed proposal. The expectation was that there would be at least one gamma ray burst satellite in the pre-selection. NASA had set up a special selection committee especially to judge gamma ray burst satellite proposals. Again, there was talk of a neck-and-neck race between the proposals of Gehrels and Haley, but in January 1999 the prize went to Gehrels's Goddard project. In the shortest possible time they had to submit a very detailed proposal and the final selection would be made in October.

Swift will be an American–British–Italian satellite weighing about 1300 kilograms, scheduled for launch in the summer of 2003. From a low orbit around the earth it will be constantly on the alert for gamma ray bursts. The Burst Alert Telescope (BAT) on board the satellite will be built by the Goddard Space Flight Center. BAT is five times as sensitive as the BATSE detector that was on board the Compton observatory and will be able to hold a sixth of the entire sky under constant surveillance. The gamma detectors will measure not only the brightness evolution of a gamma ray burst but will also record the gamma ray spectrum. As soon as a new burst is detected (Gehrels expects that there will be about 300 per year), two telescopes will be aimed at the burst: the X-ray Telescope (XRT) and the Utraviolet and Optical

Telescope (UVOT). Both telescopes are being built by Pennsylvania State University in cooperation with British and Italian institutes.

Swift will provide burst positions within 12 seconds with an accuracy of a couple of arcminutes and these positions will be automatically sent to observatories on earth. The Swift observations at X-ray and ultraviolet wavelengths and visible light begin less than 50 seconds after the actual burst and 90 seconds after the explosion the sky position will be known with an accuracy of less than an arcsecond. Besides taking brightness measurements for optical bursts and afterglows – UVOT can detect objects to 24th magnitude – spectroscopic measurements will also be made so that the redshift and the distance of the gamma ray burst can be ascertained.

On Tuesday September 14, 1999, Gehrels and his colleagues made their final presentation to the NASA selection committee. 'A very important happening', says Gehrels. 'Each of the five project teams chosen in the preselection is given 45 minutes to present their proposals. The committee then usually goes behind closed doors to deliberate and makes its decision the same day. After the choice is presented to the Congress and the necessary bureaucratic rigamarole is taken care of, the winning groups are called, usually within two days.'

The Swift song, written by Padi Boyd and performed *a cappella* by a group of Goddard astrophysicists who call themselves *The Chromatics*, was not sung at the presentation in NASA Headquarters. No doubt, such frivolity would have been out of place, even though the top brass at NASA knew about the song. After all, the text is a nice and clear introduction to an ambitious project:

> Swift is the satellite that swings/Onto those brightly bursting things/To grab the multiwavelength answer of what makes them glow/.../Swift is designed to catch a burst of gamma rays on the fly.[1]

[1] The full text of the song can be found on http://www.pagecreations.com/astrocappella/swift.html.

The days following the 14th of September were tense. By Friday the 17th, Gehrels still hadn't heard anything. He had just about given up hope when on his computer screen there came an e-mail from NASA headquarters saying that due to unexpected budget problems in the organization, the MIDEX decision was being put on hold. It was actually not beyond possibility that no selections would be made at all.

A month later, on October 12th, the sudden crisis was finally under control and Gehrels heard that a final decision would be made that week. 'It was an anxious time,' he says. 'You only get called if your proposal is selected. You could sit the whole day waiting in vain for the telephone to ring and then read in a NASA press release the next day that your competitor was chosen.' Rumors spread rapidly among scientists, but NASA headquarters did their work with the utmost care; there were no leaks about the decision that was already a month old.

On Thursday the 14th of October, a few days before he was going to the fifth Gamma Ray Burst Symposium in Huntsville, Gehrels was talking with his colleague, Steve Ritz. 'Suddenly the telephone rang. Normally I don't answer it but my secretary was not there and I had a feeling', he says. 'Just let me take this call', he said to Ritz. 'Congratulations', came a voice from the other side of the line. 'Swift is going through.'[2]

Gehrels was in seventh heaven, but he was not to say anything for one day except to his group. 'Of course I called Nick White immediately,' he says. 'And I told the news to my wife and my father.' Later in the day when White was in a meeting of an X-ray working group, everybody asked if he had heard anything but he had to remain noncommittal. That was until someone came running into the room and yelled, 'Have you heard the good news?' The astronomical rumor machine was back in action.

The choice of Swift was also big news at the Huntsville symposium. The future for gamma ray burst research suddenly looked very rosy. The

[2] The second MIDEX project that got the green light was FAME (Full-sky Astronomic Mapping Explorer), a satellite to measure accurately positions and motions of 40 million stars in the Milk Way galaxy.

Compton observatory was still working at full speed; BeppoSAX was delivering an accurate burst position about once a month; the HETE 2 launch was planned for January 28, 2000, and in 2003 Swift would add a whole new chapter to an adventure that had started 30 years before. Compton had charted the distribution of gamma ray bursts in the sky; BeppoSAX had played its vital part in the discovery of afterglows and the determination of distances; HETE 2 would get better statistics and hopefully shed new light on the still puzzling *short* bursts; and the detailed research by Swift is expected to give a crucial clue to the astrophysical processes that lead to the cosmic super-explosions.

But barely had the Swift celebratory joy subsided before set-backs showed up on all fronts. In December one of the gyroscopes on board the Compton observatory stopped working and from the beginning it was known that NASA wanted to deorbit the satellite safely as soon as possible. There was even talk that the Compton would be brought down as soon as March 2000. Around the turn of the year BeppoSAX began having a different kind of trouble. The Italian space agency ASI (Agenzia Spaziale Italiana) wanted to make deep cuts in the project budget and even considered taking the successful satellite out of service in April 2000. Fortunately that didn't happen but ASI did decide, a few months later, to drastically cut personnel costs by no longer staffing the ground station in Rome on a 7 day basis, with the result that there would be no fast follow-up observations of gamma ray bursts if they happened to occur on the weekend.

To add insult to injury, the launching of HETE 2 which was supposed to take place in the beginning of January, was delayed by 9 months. Not that there were any more problems with the Pegasus booster rocket; it was simply because NASA wanted, at the last minute, to run another round of tests. HETE 2 was to be the first scientific satellite to be launched in the year 2000, and after all the negative publicity around the two unsuccessful Mars probes, NASA knew that the eyes of the press, the public and the politicians would be on them. For safety's sake it was considered prudent to give the satellite an extra all-encompassing test. For that, HETE 2 had to be partially dismantled with all the risks that entails. It is always worrisome that even in the testing things can go wrong, as was shown at the end of March, with the HESSI satellite (one of the two SMEX projects that had been selected in 1997). It was subjected to a vibration test that, by accident, was ten times more violent than had been intended, with the result that the scheduled launch had to be delayed from July 2000 to late 2001.

Eventually, HETE 2 was successfully launched on 9 October, 2000, and after a few months of calibration, the satellite observed its first gamma ray burst on 13 February, 2001. The planned operational lifetime of HETE 2 is two years, but given the fact that BeppoSAX will probably be shut down in the spring of 2002, it looks likely that extra budget will be available to keep HETE 2 in working order until Swift is launched in the summer of 2003.

However, there are several other satellite projects in the offing for the near future. The European gamma ray satellite Integral, to be launched in 2002 or 2003, was not designed specifically to observe gamma ray bursts but it will be able to detect them. The gamma ray telescope on Integral will primarily be used to study steady sources of cosmic gamma rays, such as active galaxies, supernova remnants and pulsars. But should it happen that a gamma ray burst appears in the relatively big field of view of the telescope, its position can be determined with an accuracy of a couple of arcminutes. Integral also has sensitive spectrometers aboard as well as telescopes for X-ray and visible light.

Some three years after Integral, NASA plans to launch a successor to the Compton observatory: the Gamma-ray Large Area Space Telescope (GLAST). GLAST is in fact an improved version of the EGRET detector aboard the Compton (EGRET stands for Energetic Gamma Ray Experiment Telescope), an instrument that is designed to study the most energetic gamma rays. The sensitive instrumentation of GLAST is expected to detect several hundred gamma ray bursts per year and to determine, at least once a month, a very detailed gamma ray spectrum of a burst.

Integral is a European satellite planned to do research on cosmic sources of gamma radiation. Integral will also carry out observations of gamma ray bursts.

In addition to these two hefty programs – Integral and GLAST are big, expensive satellites – there will also be a number of smaller satellites launched in the coming years to study high-energy radiation. It is hoped that these will also contribute to a better understanding of the real nature of gamma ray bursts. One of the small satellites is AGILE, an Italian satellite which, in a sense, is the successor to BeppoSAX and EGRET. AGILE (Astro-rivelatore Gamma a Immagini LEggero) weighs only 60 kilograms and as far as sensitivity is concerned, it falls somewhere between EGRET and GLAST. The launch is currently planned for 2002.

And plans for the distant future? To be on top of the situation, you

The Gamma-ray Large Area Space Telescope (GLAST) is an American satellite that is planned to serve as the successor to the Compton Gamma Ray Observatory.

have to look ahead; and within the world of space research it means that after the success of project 1, and even before the launch of project 2, you have to begin planning project 3. On the initiative of Jerry Fishman, gamma ray burst researchers are already thinking about the successor to Swift. For the time being it is referred to as the Next Generation Gamma Ray Burst Observatory. It should be a big international project that will possibly consist of a number of satellites. However, Fishman will delay the design of the Next Generation observatory until the first results of Swift are in, so that the latest state of the art can be taken into consideration.

According to Neil Gehrels it is certainly not at all too soon to think about such a future project. 'For the Compton Gamma Ray Observatory, 20 years passed from its conception to its launching,' he says. 'The situation today has gotten somewhat better. Integral was proposed in 1990 and will fly in 2002; GLAST was proposed in 1997 and the launch is planned for 2005. But it is always a question of making plans and hanging in.' The new observatory will probably not be launched until about 2010.

Discovery, exploration, research. These are the three phases through which the exploitation of a new scientific discipline comes to fruition.

In the discovery phase there is an unexpected encounter with an unknown phenomenon. In the exploration phase the new territory is surveyed and mapped, and the most important problems are inventoried. And finally, the research phase is the time during which all the work is directed towards solving those problems. This is the period that is never really finished because every new answer brings new questions.

For gamma ray bursts the discovery phase is history. Many of the astronomers who are now actively involved in gamma ray burst research were still in elementary school when the primitive scintillators on the Vela satellites at the end of the 1960s first detected the powerful bursts of gamma rays from cosmic super-explosions. And the exploration phase heralded in by the satellites of Inter-Planetary Network also lies, for the most part, behind us. It reached its high point in 1997 with the discovery of afterglows, the determination of redshifts and distances, and the first insights into the physical processes that underlie the catastrophic explosions.

After the pioneering work of Compton and BeppoSAX, along with the hectic chase for afterglows by those indomitable earth-bound astronomers, it is now time for specialized missions such as Swift and the Next Generation Gamma Ray Burst Observatory. Astronomers have dedicated themselves to getting to the bottom of the puzzle of cosmic gamma ray bursts. They are determined to discover just what those phenomena in nature are that can produce such unconscionable amounts of energy.

But without a doubt, future research on gamma ray bursts will present an array of new questions. Not just about the explosions themselves – how they are caused, what role they play in the evolution of the universe, and what they teach us about the fundamental laws of nature – but also about their relationship to life on earth. It is interesting to contemplate that the terminal explosions of massive stars not only are bound up with the formation of life, but they also pose an intimidating threat to the continuation of that life they helped to create.

16 War and peace

'I can't tell you just now what the moral of that is, but I shall remember it in a bit.'

'Perhaps it hasn't one', 'Alice ventured to remark.

'Tut, tut, child', said the Duchess. 'Everything's got a moral, if only you can find it.'

Heaven and earth are two different worlds. Earth, the dwelling place of mankind, is a world of unpredictable change, of death and destruction, of transient imperfection. But heaven is constant, the seat of the divine and the supernatural, perfect in its order and uniformity, incorruptible in its eternity. It is difficult to think of a greater contrast. Heaven and earth, cosmos and chaos, purity and pollution.

Of course, this picture of cosmic constancy is pure sham. Stars die, galaxies evaporate, the universe expands. Everything is in constant motion; the cosmos evolves and nothing is as it was. For us on earth this cosmic flux can be difficult to perceive. On the scale of the universe, all of humanity has existed for less than a day and the lifetime of a person passes in the cosmic wink of an eye. We see no more of the life of the cosmos than a fleeting moment frozen on a piece of film, and we hastily conclude that the universe is eternal and unchanging.

Despite this gulf between heaven and earth, we have always sought to see a connection between the two realms. We once thought that the sky is the home of the gods where the sun, the moon and the planets are seen as the personifications of supernatural powers. We believed that the movements of those mysterious celestial bodies would give us some insight into the motives of the gods themselves and the control they wielded over our lives. It was believed that those who could read the language of the heavens could predict the future of the earth.

Astrology, the study of heavenly bodies and their influence on earthly happenings, dates back thousands of years. And even though a relationship with the kingdom of the gods is no longer essential to most people, astrology still enjoys a bonanza of popularity in the twenty-first century. The idea of a caring and intimate relationship between that extended, unreachable and incomprehensible world of the universe and our everyday life here on earth is a source of comfort to many; it helps us believe that we are not an insignificant, fleeting phenomenon in a big and indifferent universe; we are a part of the greater whole. We count.

It has only been in the past several hundred years that astronomers disengaged themselves from astrology. Tycho and Kepler still made up horoscopes, but by the end of the seventeenth century it was clear that the movements of the heavenly bodies were controlled by gravity rather than by the capriciousness of the gods. While astrology kept itself busy with its fancied supernatural influences of heaven on earth, astronomy directed its research to the measurable, physical powers of the cosmos. It is easy to figure out that the tremendous distances of the planets make it impossible for them to exert an influence on the earth. If Pluto suddenly came to stand on the other side of the solar system or if Mars all at once disappeared from the stage, it wouldn't make any difference. Apart from ebb and flow that is spurred by the tidal forces of the sun and the moon on the oceans, the earth is untouched by the heavens.

The break between astronomy and astrology is complete; the cleft between the two cannot be bridged, and not too many people are upset about it. Astronomers have a revulsion to pseudo-science and they shun the astrological concept of cosmic influences on earth. Maybe this adamant attitude against astrology also led astronomers to disregard the actual physical influences of the cosmos on the earth. But these do indeed exist. The earth is not an isolated planet; the cosmos forever brands its inescapable mark upon it. Life and death here on earth are inseparable from the events that take place in the universe.

Two hundred years ago the prestigious French *Académie des*

Sciences absolutely refused to listen to anything about the hypothesis that meteorites are stones from outer space. This had been suggested by the German physicist, Ernst Chladni in 1794. Only after a number of people in l'Aigle, on April 26, 1803, actually saw glowing stones falling from the sky and a French astronomer was called in to study the phenomenon, could it no longer be denied: the sky was falling down on the earth.

Cosmic projectiles may perhaps pose the greatest 'heavenly' threat to our terrestrial domain. Small meteorites that hit the surface of the earth almost every day are quite harmless (although there have been instances of houses, cars, animals and even people being hit by meteorites), but every once in a while our planet collides with a larger fragment such as an asteroid or a comet. Such heavy impacts happen very seldom but the aftermath is catastrophic.

Sixty-five million years ago the earth had such an encounter with a cosmic projectile that had a diameter of about 10 kilometers. The impact created a 200 kilometer wide crater. Billions of billions of tons of matter were hurled into the atmosphere and for weeks the earth was shrouded in darkness. Colossal floods and gigantic forest fires ravaged the planet's surface. Food chains were broken, numberless species of animals became extinct and it rang the death knell for the reign of the dinosaurs. The evolution of life on earth took a new turn as a result of a cosmic catastrophe.

According to estimates, the earth is hit on the average of about once every hundred million years by an object with a diameter of 10 kilometers or more. Smaller ones collide more frequently but their effect is less disastrous, although they could be the cause of temporary climate disturbances and the extinction of certain biological species. From the human viewpoint, 100 million years seems like a long time but cosmologically speaking, it's a drop in the bucket: less than 1% of the lifetime of the universe. The earth is forever under attack.

But it is not only the crashing of asteroids and comets that can be fatal for life on earth. Gravitational disturbances in the solar system can throw the axis of the earth out of balance. Periodic changes in the earth's orbit cause ice ages. Unexpected variations in the luminosity of the sun lead to climate oscillations. And in less than a billion years the sun will get so large and so bright that all the oceans on earth will evaporate.

Outside the solar system too, there are dangers lying in wait. No one knows what will happen if the sun in its orbit around the center of the Milky Way galaxy happens to pass through a dense cloud of interstellar molecules. Stars that pass a short distance away could cause disturbances in the comet cloud that rings the sun, triggering a bombardment of the skies. If a supernova exploded within a distance of 100 light years, so many high-energy particles would bombard the earth that the magnetosphere could no longer offer adequate protection and radiation sickness would run rampant. And if a gamma ray burst happened in our part of the Milky Way galaxy, it could mean the end of life on earth.

As we know, gamma ray bursts are the most powerful explosions in the universe – the most energetic phenomena in nature. They are more luminous than a whole galaxy of hundreds of billions of stars and in a fraction of a second they produce more energy than the sun does in 10 billion years. Every day, somewhere in the universe such a cosmic super explosion takes place, and sooner or later it will be the turn of our Milky Way galaxy. Nobody knows exactly how often a gamma ray burst is produced in galaxies such as ours; perhaps not more than one in 100 million years. But the fact is it could happen today or tomorrow; the catastrophic death of a super-massive star doesn't give much warning in advance of its coming.

A gamma ray burst at a distance of less than 3000 light years away has at least as much destructive power as the smashing into the earth of a 10 kilometer comet. The enormous expulsion of high-energy gamma rays might be held back by the earth's atmosphere but the result would be the formation of large amounts of nitric oxides. The transparent atmosphere would turn into a dark layer of smog and the sunlight would no longer reach the surface of the earth. The cancer-inducing ultraviolet radiation from the sun would have free play because the chemical reactions that produce the nitric oxides would also break down the protective ozone layer. Plants and animals would die, the ecosystem would be disrupted and people would have to hide away for many, many years.

Even worse than the actual gamma ray burst is the bombardment of high-energy electrically charged particles. For several weeks the earth would be assailed by a flood of cosmic rays. The particles with energies of trillions of electron volts produce a torrent of secondary particles in the atmosphere which would easily reach the earth's surface and go down to the bottom of the sea. According to Peter Leonard and Jerry Bonnell from NASA's Goddard Space Flight Center, who described this doomsday scenario in the American monthly *Sky & Telescope*, all life on earth would be exposed to a level of radiation a hundred times higher than a lethal dose. Moreover, as a result of the high-energy particles, radioactive material would spread itself over the entire earth.

Of course, the chance is infinitesimally small that a gamma ray burst would occur less than 3,000 light years away from the earth and the sun. But the sun is not the only star in the Milky Way galaxy with a planetary system. Since 1995 there have been a number of 'exoplanets' discovered, i.e. planets around other stars. Everything seems to point to the possibility that at least one in twenty of the sun-like stars in the universe are accompanied by one or more planets, and it would be a safe bet that each gamma ray burst irradiates and scorches some planetary systems somewhere in the universe. Unless the earth is the only inhabited planet in the universe – which according to most astronomers is highly unlikely – in the past billion years countless numbers of inhabited planets must have been completely sterilized by the cosmic violence of gamma ray bursts.

James Annis, an astrophysicist at the Fermi National Accelerator Laboratory (Fermilab) in Batavia, Illinois, even suggested that gamma ray bursts offer the solution to the 50-year-old Fermi paradox. Enrico Fermi, the Italian particle physicist and Nobel prize winner for whom the Fermilab was named, pondered over the fact that we have not been able to find any indication for intelligent life in the cosmos. If extraterrestrial civilizations exist, so reasoned Fermi, they would be able to cross and colonize the Milky Way galaxy relatively quickly. Even if the existence of intelligent life is a very rare occurrence, you would still expect that the few civilizations that do exist would spread themselves over the Milky Way galaxy. 'Where are they?' Fermi asked himself aloud.

Annis thinks that the solution to the Fermi paradox lies in the fact that gamma ray bursts regularly sterilize a large part of every galaxy in the universe. Long before a possible civilization could spread itself over a complete galaxy, let alone migrate to other galaxies, there was a catastrophic gamma ray burst nearby that completely destroyed all life.

It is a sobering thought that millions of years of biological evolution can be wiped out in one fell swoop because of the birth of a black hole. And perhaps this depressing scenario is also waiting to play itself out on earth. Who knows if earth's evolutionary clock will shortly be turned back and the adventure will start all over again with single-celled organisms in the ocean. Or then again, perhaps the earth will remain a bare and lifeless planet while in some distant place in the universe life will take a new foothold.

Most cosmic catastrophes are too big and energetic to be stopped. We cannot hold back the swelling up of the sun; we have no influence over the shape of the orbit of the earth, and if a large comet appeared from nowhere and hurtled in our direction, we have no defense. With asteroids that orbit in the inner parts of the solar system, the situation is a bit different. In principle, it is possible to track all potentially threatening objects and map out their orbits so that a collision course could be anticipated well ahead of time. With the help of rocket motors or nuclear explosions the orbit of the projectile could be somewhat deflected and altered so that the impact danger could be averted.[1] But supernova explosions and gamma ray bursts cannot be influenced by man.

The cosmos is truly a battlefield where a war is always being waged between constructive and destructive forces of nature. Atoms are fused

[1] One of the projects solely devoted to hunting for such near-earth ateroids (NEAs) is the Spacewatch project at the University of Arizona headed by Tom Gehrels, father of Neil Gehrels.

together to form molecules and organic chains, single-celled organisms evolve into more complex life forms and at least on one planet the wonder of biological evolution has led to the formation of intelligence and self-awareness. Still, at the same time, comet impacts, stellar explosions and gamma ray bursts could flatten this delicate biological framework to the ground.

But are we really talking about a war? Could we do without these cosmic catastrophes? Most likely not. Comet impacts are able to decimate life on earth, but several billion years ago they were responsible for delivering to the earth those very organic building blocks necessary for the construction of life. Comets appear to contain great quantities of organic molecules and in the early solar system they were of cardinal importance in 'impregnating' our planet. Without comet impacts there would be no life on earth.

And this is also true for supernovae. Of course, the radiation from a supernova explosion would be fatal for earthly life if it were that close, but without stellar evolution and supernova explosions no heavy elements would have come into the universe, and so again, there would have been no life. We are made from the stuff of stars that was blown into space billions of years ago by energetic stellar explosions and we are an inexorable part of the cosmic cycle.

Even the far-reaching climate fluctuations that the earth has undergone have their positive side. The fact that they have led to the extinction of a great number of biological species probably had a stimulating influence on the pace of evolution. Each catastrophe has been followed, in not too long a time, by an enormous differentiation of new species to fill up all the ecological niches that came free.

And what about gamma ray bursts – what are they good for? However these cosmic super explosions came to be, it is almost certain that they are the birth cries of black holes. The major part of the exploding object – a super-massive, fast-spinning giant star, or a merger of two neutron stars – disappears straight away into the black hole. So, as far as we know, gamma ray bursts do not contribute any significant quantity of heavy elements to enrich interstellar space. Nor is it possible to say whether or not these violent outbursts of energy could have had a role in the formation of life.

But perhaps it is too soon to try to answer such questions. The research on gamma ray bursts is still in its infancy. Many astronomers who have devoted their energies for years to the study of the sudden super-explosions have just started to unravel the mystery by tracking radio waves, X-rays and optical afterglows. *Flash!* is a book in which the end has not yet been written; no one knows what surprising turns the adventure will take and what astonishing facts will yet be brought to light. Perhaps gamma ray bursts do play a role in the formation of stars, of planets, and of life itself.

Does this imply that we must make peace with the threats from the universe? Must we meekly stand by and accept that the happenings in the cosmos could suddenly bring an end to all existence? And aren't these thoughts terribly depressing?

At first, it would seem so. But the threats from the cosmos are no more terrifying or more oppressive than the threats here on earth. Life is a short interlude between birth and death which can at any moment be disordered by outside influences. We have learned to live with most of these dangers. We accept them because they are part of the same world we live in. It is a question of violent nature and its victims, of predator and prey, of viruses and body cells. On one side there is an ongoing war that can bring death but on the other side there is the embracing peace of a subtle natural balance between the forces, organisms and processes, each of which forms an essential part of our living world.

That living world reaches far beyond the boundaries of our planet. We are less familiar with a comet impact than we are with earthquakes, but both natural calamities are inextricably tied in with the ecology of the solar system; without cosmic impacts or plate tectonics we would not be here. And supernova explosions may be more incomprehensible or even less relevant than flooding, but both phenomena are intimately connected with two vital ingredients of life: heavy elements and water.

The world in which we live is inextricably bound up with the cosmos where new stars and planets are born, but also where catastrophic collisions and explosions take place.

Heaven and earth are one. We are part of the same universe where galaxies have formed, where clouds of gas contract to create new stars and where planets are sculpted out of dust particles that have clumped together; the same universe where heavenly bodies collide with each other, stars explode and matter is sucked into unfathomable black holes. This one cosmos is the world in which we, like the stars, are born, and live and die, in which there is a balance between creation and destruction. Without this balance between war and peace the universe could not exist and we would not be here to stand in wonderment of the extravaganza nature concocts for us.

Gamma ray bursts are incomprehensible, puzzling, mysterious. But they belong to the same universe as the astronomers who are trying to unlock their secret. Other temperatures and energies prevail but the elementary particles in the relativistically expanding fireball are identical to the building blocks of the human body. The gravity that is so unimaginably strong that it can swallow a complete star in a fraction of a second is the same gravity to which we owe our sense of balance. The laws of nature that govern the extreme circumstances of distant realms and the time-enshrouded past are the same laws of nature at work in the more serene here and now. It is one world, one cosmos, one universe.

If there is a moral to be found here, it is that we are all part of a fantastic miracle.

Glossary

"What do you mean by that?" said the Caterpillar sternly.
"Explain yourself!"

absorption line A thin dark line that is seen in the spectrum of a celestial body caused by the absorption of light at a specific wavelength by matter that is found between the light source and the observer.

accretion disk A rapidly rotating disk of matter around a compact celestial body (such as a neutron star or a black hole) that attracts gas and dust from its surroundings.

afterglow A light phenomenon, decreasing in luminosity, that can sometimes be observed at different wavelengths for several hours, days or weeks after a gamma ray burst.

Andromeda galaxy The closest large galaxy comparable to the Milky Way galaxy.

anomalous X-ray pulsar A young neutron star that sends out especially low-frequency, short pulsations of X-rays.

antimatter An all-encompassing name for particles that have the same mass as regular particles of matter but have an opposite electrical charge.

arcminute One sixtieth of a degree; the angle which an American dime subtends at a distance of about 60 meters.

arcsecond One sixtieth of an arcminute; the angle which an American dime subtends at a distance of about 3.6 kilometers.

asteroid A small rock-like celestial body in orbit around the sun.

beaming A phenomenon in which radiation is not sent out isotropically, but rather in two oppositely directed jets.

big bang The popular description of the birth of the universe in an enormous explosion in which space, time, energy and matter originated.

BL Lacertae-object An active galaxy comparable to a quasar and named for the prototype BL Lacertae.

black hole A region in the universe in which the strength of the gravitational field is so strong that not even light can escape from it.

brown dwarf A star whose mass is not great enough to produce internal nuclear fusion.

clean room A dust-free room in which satellites are built and tested.

cluster A collection of galaxies that are held together by mutual gravity.

collapsar An extremely massive, rapidly rotating star that collapses into a black hole at the end of its life.

comet A small, icy celestial body that normally describes an elongated orbit around the sun.

constellation An apparent grouping of relatively bright stars in the heavens.

cosmic background radiation Microwave radiation which is uniformly distributed in the universe and is observed as the cooled remnant of the radiation produced during the big bang.

cosmic rays High-energy particles from the universe, consisting of energetic electrons, protons and atomic nuclei.

cosmology The area of astronomy dealing with the birth, evolution and structure of the universe as a whole.

dark matter Matter that does not emit any radiation but whose gravitational effect is felt.

declination The angular distance of a celestial body from the celestial equator, equivalent to geographic latitude on the earth.

degree A ninetieth of a right angle; the angle an American dime subtends at a distance of about a meter.

doppler effect A phenomenon in which the wavelength of the radiation is dependent upon the movement of the radiation source relative to the observer.

electromagnetic spectrum An all-encompassing name for all forms of radiation.

electron An elementary particle with a very small mass and a negative electrical charge.

electron volt The energy gained by an electron when it passes through an electrical potential of one volt.

elementary particle The fundamental building block of nature.

emission line A thin, bright line in the spectrum of a celestial body caused by radiation that is sent out at a specific wavelength.

erg A unit of energy (1 erg = 10^{-7} joule).

error box A small area in the sky which is the possible position of a celestial body, allowing for possible uncertainties in measurements of the position.

extragalactic Located in or originating from an area outside of the Milky Way galaxy.

frequency Number of vibrations per second.

galactic Located in or originating from an area inside of the Milky Way galaxy.

galaxy A large number of stars bound together by their mutual gravity.

gamma radiation Electromagnetic radiation with a wavelength shorter than a hundredth of a nanometer.

gamma ray burst A short, powerful burst of cosmic gamma radiation.

gamma ray burst position The position in the sky of a gamma ray burst, usually given in right ascension and declination.

gamma ray detector An instrument to detect gamma rays.

gauss The unit measuring the strength of the magnetic field; the strength of the magnetic field on the surface of the earth is 0.6 gauss.

giant star A star whose brightness is substantially greater than that of a normal star with the same surface temperature.

gravitational collapse Collapse of the interior of a star resulting from gravitation exceeding the gas pressure.

gyroscope An instrument for stabilizing the orientation of a satellite.

high-energy astrophysics The area of astrophysics in which high-energy processes play a role.

host galaxy A galaxy in which a gamma ray burst took place.

hypernova An extremely energetic supernova.

infrared radiation Electromagnetic radiation with a wavelength between 700 nanometers and 1 millimeter.

interferometry An observational technique by which a very sharp picture can be obtained by letting two or more light beams from the same source interfere with each other.

interstellar matter Rarefied material in the space between the stars.

ion An electrically charged atom; an atom that has lost one or more electrons, or has one or more electrons too many attached to it.

isotope One of a set of species of atoms with the same number of protons in the atomic nucleus but with different numbers of neutrons.

isotropic distribution A distribution that is not dependent upon direction.

jet A rapid, narrowly directed outpouring of matter.

Large Magellanic Cloud A satellite galaxy of the Milky Way galaxy.

light curve A graphical description of the brightness of a celestial object as a function of time.

light year The distance a ray of light travels in one year: 9.46 trillion kilometers.

magnetar A young neutron star with an extremely strong magnetic field.

magnetohydrodynamics A subfield of fluid dynamics which includes the effects of magnetic fields.

magnitude The standard unit of brightness of a celestial body. A star of 6th magnitude (just barely visible to the naked eye) is 100 times dimmer than a star of 1st magnitude (the brightest stars in the sky).

Milky Way galaxy The spiral-shaped distribution of stars in which our sun is located.

millisecond pulsar A pulsar that emits more than 50 pulsations per second as a result of its rapid rotation.

nanometer One millionth of a millimeter.

neutrino An elementary particle without charge and with little or no measurable mass. Neutrinos are not a part of atoms and molecules but are extremely abundant in the universe.

neutron A nuclear particle without charge whose mass is slightly greater than that of the proton. Neutrons are only stable within atomic nuclei.

neutron star A very compact star a few times as massive as the sun but with a diameter of only about 30 kilometers; a remnant of a supernova explosion.

nuclear fusion The fusing of light atomic nuclei into heavier ones; it can occur spontaneously under very high pressure and very high temperatures.

Oort cloud An extended cloud of comets which are orbiting very slowly at a great distance from the sun.

optical burst A short burst of visible light sometimes associated with a gamma ray burst.

peak luminosity The greatest luminosity reached in a particular time interval.

photomultiplier tube A light-sensitive instrument which converts very faint light signals into clearly measurable electric currents.

photon The smallest unit of radiation.

planet A cool and relatively large celestial body orbiting a star.

planetary nebula The shell of gas blown out by an unstable giant star before it evolves into a white dwarf.

proton A nuclear particle with a positive electric charge and a mass about 1836 times that of an electron.

pulsar A neutron star that emits regular short pulses of radio waves.

quasar The bright star-like nucleus of a very distant active galaxy.

radiation pressure The pressure produced by electromagnetic radiation.

radio afterglow The afterglow of a gamma ray burst at radio wavelengths.

radio telescope An instrument for observing cosmic radio waves, mostly consisting of one or more large dish antennae.

radio wave Electromagnetic radiation with a wavelength greater than several millimeters.

redshift The increase in wavelength of radiation emitted by a source moving away from the observer.

relativity theory A theory of the structure of space and time and the nature of gravitation.

right ascension The sky coordinate equivalent to geographical longitude on the earth.

satellite An unmanned space vehicle orbiting the earth.

scintillation Short, rapid variation in the brightness and sky position of a point-shaped celestial body.

scintillator Material in which weak light flashes are produced by penetrating gamma ray photons or high-energy particles.

sky position The position of a celestial body in the sky, usually given in right ascension and declination.

soft gamma repeater A source of periodic but irregular outbursts of relatively soft gamma rays.

solar flare A powerful explosion on the sun which produces radio radiation, X-rays and gamma rays.

space probe An unmanned space vehicle traversing the solar system.

spectral line A bright or dark line in the spectrum of a celestial body.

spectrometer An instrument that records the spectrum of a celestial body.

spectroscopy The field of astronomy in which research is done on the spectra of celestial bodies.

spectrum The graphic display of the brightness of a celestial object as a function of wavelength.

speed of light The speed of transmission of electromagnetic radiation in a vacuum: 299,792,458 kilometers per second.

spiral arm The region in a spiral-shaped galaxy which contains the brightest stars.

star A large gaseous celestial body that emits energy produced by nuclear fusion processes in its interior.

star-forming region Interstellar gas and dust cloud in which new stars are formed.

supercluster A large cluster of galaxies.

supernova A gigantic explosion that comes at the end of the life of a massive star.

supernova remnant The expanding gaseous remains of a supernova explosion.

telescope An instrument to observe the electromagnetic radiation coming from the universe.

time resolution The accuracy with which the timing of an event can be determined.

ultraviolet radiation Electromagnetic radiation with a wavelength between 10 and 400 nanometers.

visible light Electromagnetic radiation with a wavelength between 400 and 700 nanometers.

wavelength The distance between the adjacent crests of an advancing wave.

white dwarf A small, compact star with a high surface temperature.

wide-angle camera A camera with a relatively large field of view (in astronomy: greater than several degrees).

X-rays Electromagnetic radiation with a wavelength between 0.01 and 10 nanometers.

X-ray afterglow The afterglow of a gamma ray burst at X-ray wavelengths.

X-ray binary A binary star system that emits X-rays mostly as a result of matter transfer.

X-ray burster A star (most often an X-ray binary) which emits a large amount of X-rays non-periodically.

Acronyms

'You seem very clever at explaining words, Sir', said Alice.
'Would you kindly tell me the meaning of the poem called "Jabberwocky"?'

AAS American Astronomical Society
AAT Anglo-Australian Telescope
AGILE Astro-rivelatore Gamma a Immagini Leggero
ANS Astronomical Netherlands Satellite
ApJ *Astrophysical Journal*
ASCA Advanced Satellite for Cosmology and Astrophysics
ASI Agenzia Spaziale Italiana
AXP Anomalous X-ray Pulsar
BACODINE BAtse COordinates DIstribution NEtwork
BASIS Burst Arc Second Imaging and Spectroscopy
BAT Burst Alert Telescope
BATSE Burst And Transient Source Experiment
BOLT Broadband Observatory for the Localization of Transients
Caltech California Institute of Technology
CCD Charge Coupled Device
CERN Centre Européen pour la Recherche Nucléaire
CESR Centre d'Étude Spatiale des Rayonnements
CGRO Compton Gamma Ray Observatory
COMIS COded Mask Imaging Spectrometer
COMPTEL COMPton TELescope
COPUOS Committee On the Peaceful Uses of Outer Space
CXO Chandra X-ray Observatory
DSS Digitized Sky Survey
EGRET Energetic Gamma Ray Experiment Telescope
ESA European Space Agency

ESO European Southern Observatory
ESRO European Space Research Organization
ETC Explosive Transient Camera
FAME Full-sky Astrometric Mapping Explorer
GALEX Galactic Evolution Explorer
GCN GRB Coordinates Network
GERSE Gamma-ray Energetic Radiation Spectroscopy Experiment
GLAST Gamma-ray Large Area Space Telescope
GRB Gamma Ray Burst
GRBM Gamma Ray Burst Monitor
GRO Gamma Ray Observatory
GROCSE Gamma Ray Optical Counterpart Search Experiment
GSFC Goddard Space Flight Center
HESSI High Energy Solar Spectroscopic Imager
HETE High Energy Transient Explorer
HST Hubble Space Telescope
IAS Istituto di Astrofisica Spaziale
IAU International Astronomical Union
IFC Istituto di Fisica Cosmica
IMP Interplanetary Monitoring Platform
Integral International Gamma Ray Astrophysics Laboratory
IPN Inter-Planetary Network
IRAS Infra-Red Astronomical Satellite
ISEE International Sun–Earth Explorer
KPNO Kitt Peak National Observatory
LAEFF Laboratorio de Astrofísica Espacial y Física Fundamental
LANL Los Alamos National Laboratory
LECS Low Energy Concentrator Spectrometer
LLNL Lawrence Livermore National Laboratory
LOTIS Livermore Optical Transient Imaging System
MACHO MAssive Compact Halo Object

MECS Medium Energy Concentrator Spectrometer
MHD Magnetohydrodynamica
MIDEX Medium-class Explorer
MIT Massachusetts Institute of Technology
MNRAS *Monthly Notices of the Royal Astronomical Society*
MFSC Marshall Space Flight Center
NAE Nuclear Astrophysics Explorer
NASA National Aeronautics and Space Administration
NGGO Next Generation GRB Observatory
NIVR Nederlands Instituut voor Vliegtuigontwikkeling en Ruimtevaart (Netherlands Agency for Aerospace Programs)
NOAO National Optical Astronomy Observatories
NRAO National Radio Astronomy Observatory
NTT New Technology Telescope
ROSAT RÖntgen SATellite
ROTSE Robotic Optical Transient Search Experiment
RXTE Rossi X-ray Timing Explorer
SAC Satélite de Aplicaciones Científicas
SAX Satellite per Astronomia X
SDI Strategic Defense Initiative
SGR Soft Gamma Repeater
SIRTF Space Infra-Red Telescope Facility
SMEX Small Explorer
SMM Solar Maximum Mission
SN Supernova
SRON Space Research Organization Netherlands
TDRSS Tracking and Data Relay Satellite System
TEM Transient Event Monitor
TeSRE Istituto Tecnologie e Studio delle Radiazione Extraterrestri
TIXTE TIming X-ray Transient Explorer
UVOT Ultra-Violet Optical Telescope
VLA Very Large Array
VLT Very Large Telescopc

WFC Wide Field Camera
WHT William Herschel Telescope
WSRT Westerbork Synthesis Radio Telescope
XMM X-ray Multi Mirror
XRT X-Ray Telescope

Sources and literature

'I think I should understand that better', Alice said very politely, 'if I had it written down: but I can't quite follow it as you say it'.

The citations at the beginning of each chapter come from *Alice's Adventures in Wonderland* and *Through the Looking-Glass* by Lewis Carroll.

1. The sky watchers of Los Alamos

Chapter 1 is, for the most part, based upon conversations with Ed Fenimore and Stirling Colgate (November 17, 1999, in Los Alamos, New Mexico), Ray Klebesadel (November 20, 1999, in Waukesha, Wisconsin) and Kevin Hurley (January 7, 2000, in Berkeley, California). The discovery of gamma ray bursts by Klebesadel, Strong and Olson in 1973 was written up in *The Astrophysical Journal*, vol. 182, p. L85. A popular version written by Strong and Klebesadel, entitled, 'Cosmic gamma-ray bursts', appeared in October 1976 in *Scientific American* (p. 66).

2. The bat mystery

Chapter 2 is based upon a visit to NASA's Marshall Space Flight Center in Huntsville, Alabama, on January 11, 2000, and conversations with Gerald Fishman and Charles Meegan on the 10th and 11th of January. The first balloon observations of gamma ray bursts were described in *The Astrophysical Journal*, vol. 291, p. 479 ('The frequency of weak gamma ray bursts', by Meegan, Fishman and Wilson). BATSE's discovery of the isotropic distribution of gamma ray bursts appeared in *Nature*, January 9,1992, p. 143 (Meegan *et al.*). General articles about the first results from the Compton Gamma Ray Observatory are, 'Probing the gamma ray sky' (Hurley), *Sky & Telescope*, December 1992, p. 631, and 'The gamma ray cosmos' (Chupp), *Science*, December 18,1992, p. 1894. The website for the CGRO is http://cossc.gsfc.nasa.gov/cossc/index.html.

3. A duel over distance

Chapter 3 is, for the most part, based upon conversations with Bohdan Paczyński (October 23, 1999, in Princeton, New Jersey) and Donald Lamb (November 19, 1999, Chicago, Illinois). Both astronomers contributed to the special issue about 'The Great Debate' of *Publications of the Astronomical Society of the Pacific* (vol. 107, no. 718, December 1995): 'The distance scale to gamma ray bursts', (Lamb, p. 1152) and 'How far away are gamma ray bursters?' (Paczyński, p. 1167). Earlier publications concerning the distance dilemma are: 'Taking stock of gamma ray bursts' (Hartmann), *Science*, January 7, 1994, p. 47; 'The mystery that won't go away' (Meegan), *Sky & Telescope*, August, 1994, p. 28 and 'Gamma ray bursts: near or far?' (Horack), *Science*, August 26, 1994, p. 1186. A popular review of the debate is: 'Gamma ray bursters: near or far?' (Talcott), *Astronomy*, December 1995, p. 56.

4. Liras, tears and satellites

Chapter 4 is primarily based upon a visit to the BeppoSAX Science Operations Center in Rome on October 8, 1999, and conversations with Enrico Costa (October 9, 1999, in Rome) and Donald Lamb (November 19, 1999, in Chicago, Illinois). The website for BeppoSAX is http://www.sdc.asi.it/sax-main.html; information about HETE can be found at http://space.mit.edu/hete.

5. Beaming in on afterglows

Chapter 5 is based upon conversations with Hans Muller, Luigi Piro and Marco Feroci (October 8, 1999, in Rome), Enrico Costa (October 9, 1999, in Rome), John Heise and Jean in 't Zand (November 10, 1999, in Utrecht) and Dale Frail (November 16,1999, in Socorro, New Mexico). The X-ray wide field cameras of BeppoSAX are described in 'The wide field cameras onboard the BeppoSAX x-ray astronomy satellite' (Jager *et al.*), *Astronomy & Astrophysics*, April 11, 1997. The website for the wide field cameras is http://wfc.sron.nl. The discovery of X-rays from GRB 960720 is described in 'Search narrows for gamma ray bursts' (Schilling), *Science*, October 4, 1996, p. 38.The radio observations of GRB 970111 are described in 'Radio monitoring of the 1997 January 11 gamma ray burst' (Frail *et al.*), *The Astrophysical Journal*, vol. 483, p. L91 and 'Radio and optical follow-up observations and improved IPN position of GRB 970111' (Galama *et al.*), *Astrophysical Journal Letters*, vol. 493, p. 27.

6. First among equals

Chapter 6 is, for the most part, based upon conversations with Jan van Paradijs (various dates between 1997 and 1999), Enrico Costa (October 9, 1999, in Rome), Paul Groot (November 10, 1999, in Amsterdam), John Heise (November 10, 1999, in Utrecht), Kailash Sahu (November 23, 1999, in Baltimore, Maryland) and Titus Galama (January 5, 2000, in Pasadena, California). The discovery of the optical afterglow of GRB 970228 is described in 'Transient optical emission from the error box of the gamma ray burst of 28 February 1997' (van Paradijs *et al.*), *Nature*, April

17, 1997, p. 686. The discovery of the X-ray afterglow is described in 'Discovery of an x-ray afterglow associated with the gamma ray burst of February 28, 1997 (Costa *et al.*), *Nature*, June 19, 1997, p. 783. Other *Nature* publications about GRB 970228 are: Paczyński and Wijers (May 17, p. 650), Sahu *et al.* (May 29, 1997, p. 476) and Galama *et al.* (May 29, 1997, p. 479). Popular articles about the first identifications are: 'God's firecrackers' (Walker), *New Scientist*, May 31, 1997, p. 28 and 'Gamma ray bursts' (Fishman and Hartmann), *Scientific American*, July 1997, p. 34.

7. Eavesdropping on heavenly whispers

Chapter 7 is based upon a visit to the Very Large Array and a conversation with Dale Frail (November 16, 1999, in Socorro, New Mexico), and a conversation with George Djorgovski (January 5, 2000, in Pasadena, California). The website for the VLA is http://info.aoc.nrao.edu/vla/html/vlahome.shtml. The relation between soft gamma repeaters and supernova remnants is described in 'The radio nebula of the soft gamma ray repeater 1806–20' (Kulkarni *et al.*), *Nature*, March 10, 1994, p. 129. The discovery of the optical afterglow of GRB 970508 is described in 'Optical counterpart to the gamma ray burst GRB 970508' (Djorgovski *et al.*), *Nature*, June 26, 1997, p. 876 and 'Spectral constraints on the redshift of the optical counterpart to the gamma ray burst of 8 May 1997' (Metzger *et al.*), *Nature*, June 26, 1997, p. 878. The radio observations of GRB 970508 are described in 'The radio afterglow from the gamma ray burst of 8 May 1997' (Frail *et al.*), *Nature*, September 18, 1997, p. 261 and 'Position and parallax of the gamma ray burst of 8 May 1997', (Taylor *et al.*), *Nature*, September 18, 1997, p. 263. Other publications about GRB 970508 are: Castro-Tirado *et al.* (*Science*, February 13, 1998, p. 1011), Galama *et al.* (*Astrophysical Journal Letters*, vol. 497, p. 13), Galama *et al.* (*Astrophysical Journal Letters*, vol. 500, p. 97), and Galama *et al.* (*Astrophysical Journal Letters*, vol. 500, p. 101). A popular article written about the race to find afterglows is 'Speed Matters' (Zimmerman), *Astronomy*, May 2000, p. 36.

8. Thinking with the speed of light

Chapter 8 is based upon conversations with Ralph Wijers, (December 8, 1999, in Amsterdam) and Martin Rees (April 26, 2000, in Leiden). The relativistic fireball model is described by Rees and Mészáros in *The Astrophysical Journal*, vol. 476, p. 232. The physical properties (based on the observed spectral characteristics) are described in 'Physical parameters of GRB 970508 and GRB 971214 from their afterglow synchrotron emission' (Wijers and Galama), *The Astrophysical Journal*, vol. 523, p. 177.

9. Competition for the big bang

Chapter 9 is based, for the most part, on conversations with Shrinivas Kulkarni and George Djorgovski (January 5, 2000, in Pasadena, California). The observations of GRB 971214 ('Big Bang 2') are described in three articles in *Nature,* May 7, 1998: 'Identification of a host galaxy at redshift z = 3.42 for the gamma ray burst of 14 December 1997' (Kulkarni *et al.*, p. 35), 'Optical afterglow of the gamma ray burst of 14 December 1997' (Halpern *et al.*, p. 41) and 'The energetic afterglow of the gamma ray burst of 14 December 1997' (Ramaprakash *et al.*, p. 46). A commentary on the discovery appeared in the same issue of *Nature:* 'The burst, the burster and its lair' (Wijers, p. 13). A popular description of the discovery is found in 'Gamma blast from way, way back' (Glanz), *Science,* April 24, 1998, p. 514.

10. Curious connections

Chapter 10 is primarily based upon a conversation with Titus Galama (January 5, 2000, in Pasadena, California). The observations of GRB 980425 and SN 1998bw are described in three articles in *Nature*, October 15, 1998: 'Radio emission from the unusual supernova 1998bw and its association with the gamma ray burst of 25 April 1998' (Kulkarni *et al.*, p. 663), 'An unusual supernova in the error box of the gamma ray burst of 25 April 1998', (Galama *et al.*, p. 670), and 'A hypernova model for the supernova associated with the gamma ray burst of 25 April 1998' (Iwamoto *et.al.*, p. 672). A commentary on the discovery appeared in the same issue of *Nature*, 'How big do stellar explosions get?' (Baron, p. 635). For a popular description of the discovery one can read 'Supernova and gamma ray burst might have common origin' (Schilling), *Science,* June 19, 1998, p. 1836. The discovery of a possible supernova signal in the light curve of GRB 980326 is described in 'The unusual afterglow of the gamma ray burst of March 26, 1998 as evidence for a supernova connection', (Bloom *et al.*), *Nature*, September 30, 1999, p. 453. A commentary on the possible association between supernovae and gamma ray bursts is found in 'From gamma ray bursts to supernovae' (Van Paradijs), *Science*, October 22, 1999, p. 693. The work of Galama and Vreeswijk is, among others, described in 'Afterburn' (Schilling), *New Scientist,* September 25, 1999, p. 36.

11. Alchemists of the cosmos

Chapter 11 is based on a search of the literature. A recommended article about supernovae is, 'Supernova explosions in the universe' (Burrows), *Nature,* February 17, 2000, p. 727. Several popular scientific books about supernovae are *The supernova story* (Marshall), Princeton University Press, 1994, and *Supernova! The exploding star of 1987* (Goldsmith), St.Martin's Press, 1989.

12. The magnetar attraction

Chapter 12 is, for the most part, based upon a conversation with Chryssa Kouveliotou (January 11, 2000, in Huntsville, Alabama). The discovery of X-ray pulses from SGR 1806–20 is described in 'An x-ray pulsar with a superstrong magnetic field in the soft gamma ray repeater SGR 1806–20' (Kouveliotou *et al.*), *Nature*, May 21, 1998, p. 235. The discovery of magnetars is described in 'Magnetar blasts its way out of theory' (Reichhardt), *Nature*, October 8, 1998, p. 529. The outburst of SGR 1900 + 14 is described in 'A giant periodic flare from the soft gamma ray repeater SGR1900 + 14' (Hurley *et al.*), *Nature*, January 7, 1999, p. 41, and 'An outburst of relativistic particles from the soft gamma ray repeater SGR1900 + 14' (Frail *et al.*), *Nature*, March 11, 1999, p. 127. More on soft gamma repeaters and anomalous X-ray pulsars can be found in 'Astronomical odd couple? Or later egos?' (Schilling), *Science*, January 5, 2001, p. 68. A great deal of information about magnetars can be found on http://www.magnetar.com.

13. The Argus eyes of Livermore

Chapter 13 is based on visits to the ROTSE observatory in Los Alamos, New Mexico (November 17, 1999) and the LOTIS observatory in Livermore, California (January 7, 2000) and on conversations with Carl Akerlof (November 18, 1999, in Ann Arbor, Michigan), Hye-Sook Park (January 7, 2000 in Livermore, California) and Bradley Schaefer (January 13, 2000, in Atlanta, Georgia). The results of the GROCSE project are described in 'Results from Gamma Ray Optical Counterpart Search Experiment: a real time search for gamma ray burst optical counterparts' (Lee *et al.*), *The Astrophysical Journal*, vol. 482, p. L125. The observations of GRB 990123 are described in three articles in *Nature*, April 1, 1999: 'The afterglow redshift and extreme energetics of the gamma ray burst of 23 January 1999' (Kulkarni *et al.* p. 389), 'The effect of magnetic fields on gamma ray bursts inferred from multi-wavelength observations of the burst of 23 January 1999' (Galama *et al.*, p. 394) and 'Observation of contemporaneous optical radiation from a gamma ray burst' (Akerlof *et al.*, p. 400). In the same issue of *Nature* there appeared a commentary about the discovery: 'A burst caught *in flagrante*' (Mészáros, p. 368). At the same time there appeared three articles about this gamma ray burst in *Science*, March 26, 1999: 'Decay of the GRB 990123 optical afterglow: implications for the fireball model' (Castro-Tirado *et al.*, p. 2069), 'Polarimetric constraints on the optical afterglow emission from GRB 990123' (Hjorth *et al.*, p. 2073), and 'Spectroscopic limits on the distance and energy release of GRB 990123' (Andersen *et al.*, p. 2075). In the same issue of *Science* there appeared a summary of the observations (Schilling, p. 2003). The website for ROTSE is http://www.umich.edu/~rotse; the website for LOTIS is http://hubcap.clemson.edu/~ggwilli/lotis/index.html.

14. Fireworks and black holes

Chapter 14 is based upon conversations with Stan Woosley and Andrew MacFadyen (January 6, 2000, in Santa Cruz, California). A popular article about Woosley and his work is 'The supernova guru' (Irion), *Astronomy*, July 1999, p. 48. Information about the computer simulations of MacFadyen (including film clips) can be found at http://www.ucolick.org/~andrew.

15. A flashing future

Chapter 15 is based, for the most part, upon a conversation with Neil Gehrels (November 24, 1999, in Greenbelt, Maryland). A retrospective on the Compton Gamma Ray Observatory can be found in 'Compton's legacy: highlights from the Gamma Ray Observatory' (Leonard and Wanjek), *Sky & Telescope*, July 2000, p. 48. Each of the new and future scheduled gamma ray satellites has its own website. HETE 2: http://space.mit.edu/hete. Swift: http://swift.sonoma.edu. Integral: http://astro.estec.esa.nl/sa-general/projects/integral/integral.html. GLAST: http://glast.gsfc.nasa.gov. AGILE: http://www.roma2.infn.it/infn/agile.

16. War and peace

An article about the dangers of gamma ray bursts is 'Gamma ray bursts of doom' (Leonard and Bonnell), *Sky & Telescope*, February 1998, p. 28. The theory that gamma ray bursts provide the answer to the Fermi paradox is described by Annis in the *Journal of the British Interplanetary Society*, vol. 52, p. 19, and in 'Sorry, we'll be late' (Chown), *New Scientist*, January 23, 1999, p. 16.

Index

"Of course you know your ABC?" said the Red Queen.

An 'i' after a page number refers to an illustration; an 'n' refers to a footnote.